Implementing and Evaluating an Innovative Approach to Simulation Training Acquisitions

Christopher Paul, Harry J. Thie, Elaine Reardon,
Deanna Weber Prine, Laurence Smallman

Prepared for the Office of the Secretary of Defense
Approved for public release; distribution unlimited

The research described in this report was prepared for the Office of the Secretary of Defense (OSD). The research was conducted in the RAND National Defense Research Institute, a federally funded research and development center sponsored by the Office of the Secretary of Defense, the Joint Staff, the Unified Combatant Commands, the Department of the Navy, the Marine Corps, the defense agencies, and the defense Intelligence Community under Contract DASW01-01-C-0004.

Library of Congress Cataloging-in-Publication Data

Implementing and evaluating an innovative approach to simulation training acquisitions / Christopher Paul ... [et al.].
p. cm.
"MG-442."
Includes bibliographical references.
ISBN 0-8330-3903-2 (pbk. : alk. paper)
1. United States—Armed Forces—Procurement. 2. Military education—Aids and devices. 3. Synthetic training devices. I. Paul, Christopher, 1971–

UC263.I45 2006
355.5—dc22

2006005448

Published 2006 by the RAND Corporation
1776 Main Street, P.O. Box 2138, Santa Monica, CA 90407-2138
1200 South Hayes Street, Arlington, VA 22202-5050
201 North Craig Street, Suite 202, Pittsburgh, PA 15213-1516
RAND URL: http://www.rand.org/
To order RAND documents or to obtain additional information, contact
Distribution Services: Telephone: (310) 451-7002;
Fax: (310) 451-6915; Email: order@rand.org

Preface

In the wake of the failure of the Joint Simulation System (JSIMS), the Department of Defense (DoD) sought improvements to its approach to buying simulations training through a process called the Training Capabilities Analysis of Alternatives (TC AoA). The DoD has decided to move forward with a prototype of one alternative developed as part of this process—an innovative business model intended to align the financial incentives of industry participants with positive training and technology development outcomes. Known as Alternative #4 (referred to hereafter as Alt#4), the model proposes to turn what has traditionally been the acquisition of training simulators into a service acquisition (the acquisition of training) with a private sector "tool vendor" marketplace to support it. RAND was asked by the offices of the Under Secretaries of Defense (OUSD) for Personnel and Readiness (P&R) and Acquisition, Technology and Logistics (AT&L), and the Joint Staff Operational Plans and Joint Forces Development Directorate (J7), to produce an implementation and evaluation plan for a prototype of this alternative.

This report responds to that request. It presents the Alt#4 business model, compares it with other approaches for buying simulations and simulation training, reviews economic theories relevant to the model, and provides a detailed implementation and evaluation plan for a prototype.

This report should be of interest to those participating in, or charged with carrying out, the prototype of Alt#4. The analysis should also be interesting to those interested in innovative approaches to training acquisition. Moreover, the success or failure of the approach will be of general interest to the executive and legislative branches and to commercial companies as an evaluated approach to improved acquisition. No special technical expertise is required to understand the material.

Those interested in this report may also find the following RAND report of interest: Bruce Held, Kenneth P. Horn, Michael Hynes, et al., *Seeking Nontraditional Approaches to Collaborating and Partnering with Industry*, Santa Monica, Calif.: RAND Corporation, MR-1401-A, 2002.

This research was sponsored by OUSD (P&R), OUSD (AT&L), and J7. It was conducted within the Forces and Resources Policy Center of the RAND National Defense Research Institute, a federally funded research and development center sponsored by the Office of the Secretary of Defense, the Joint Staff, the Unified Combatant Commands, the Department of the Navy, the Marine Corps, the defense agencies, and the defense Intelligence Community.

For more information on RAND's Forces and Resources Policy Center, contact the Director, James Hosek. He can be reached by e-mail at James_Hosek@rand.org; by phone at 310-393-0411, extension 7183; or by mail at RAND Corporation, 1776 Main Street, Santa Monica, California 90407-2138. More information about RAND is available at www.rand.org.

Contents

Figures

Tables

Summary

In the wake of the failure of the Joint Simulation System (JSIMS), the Department of Defense (DoD) sought to improve its approach to buying training and simulations through a process called the Training Capabilities Analysis of Alternatives (TC AoA). The DoD has decided to move forward with a prototype of one alternative developed as part of this process, Alternative #4, an innovative business model that hopes to align the financial incentives of industry participants with positive training and technology development outcomes. The model proposes to turn what has traditionally been the acquisition of training simulators into a service acquisition (the acquisition of training) with a private sector "tool vendor" marketplace to support it. RAND was asked to produce an implementation and evaluation plan for this prototype. This report details the Alternative #4 (henceforth, Alt#4) business model, examining it in light of economic theory and of other business models for training and simulation acquisition. The report concludes that although Alt#4 has merit, it is not without challenges. It discusses those risks and challenges and presents detailed plans for the implementation and evaluation of a prototype of Alt#4.

The report's findings are based on analyses from two data sources:

- a review of documents, academic literature, economic theory, and publicly available information about various simulation training initiatives; and

- interviews with industry and DoD participants in the TC AoA process or with experience developing, procuring, or using simulations for training; we also interviewed personnel from both the industry and defense side of simulation training provision in the United Kingdom.

From these data, we assembled business model case summaries of relevant programs and a set of business models with various approaches to buying simulation tools or training support.

The Policy Problem

Alt#4 addresses the way the DoD has traditionally bought simulations and simulated training support. The "old" business model is characterized as being both fiscally wasteful and a hindrance to innovation because it created a system of inefficient long-term commitments to what are effectively contractor-proprietary simulation systems. The features of the old business model that Alt#4 seeks to address are detailed in Chapter Two.

At issue is how best to acquire simulations and simulation training support. Alt#4 claims to be *a* way to buy high-quality simulation training support at good value. It does not claim to be the only way to do so or even the best way under all circumstances. This report finds that Alt#4 is based on sound economic principles and has the potential to efficiently deliver high-quality training and innovation in training tools. But to reach this potential, tools should not be completely DoD-specific and training tasks should have both requirements and performance measures that can be clearly specified. A prototype of Alt#4 could demonstrate the ability of this potential to be realized in a real DoD training context.

The Alt#4 Business Model

Alt#4 proposes a twofold solution to the simulation training procurement problem: First, in areas subject to the business model, the DoD stops buying both tools *and* training support and buys only training support; second, the DoD stops buying training support with "cost-plus" contracts and starts buying it on firm-fixed-price (FFP) per training outcome contracts.

In addition, Alt#4 proposes that the DoD engage in several efforts to ensure innovation and competition in the simulation tool market:

- separate the training support and tool markets by including conflict of interest clauses in training support contracts (referred to as "untying" the markets);
- impose compliance with adopted technical standards to guarantee product operability and interoperability; and
- create a mechanism to inject seed money into the tool market to support innovation or create competition in submarkets.

The logic underlying each of these elements is to have the DoD eliminate the perverse incentive structures of previous business models and instead create incentives for the training service providers (TSPs) and tool vendors to provide responsive simulation training support using best products and practices at best value prices.

The Alt#4 business model is described in detail in Chapter Two. Figure S.1 presents a notional summary of the transformation envisioned by Alt#4. The designers of Alt#4 maintain that the biggest problems with the DoD's old way of buying simulation training stem from the vertical integration of tool and training provision in single contractors and the use of cost-plus contracts for the procurement of training support and simulation tools. Under the Alt#4 business model, the DoD would use firm-fixed-price contracts instead of cost-plus contracts and the tool market would be separated (untied) from the training service provider market. These two changes alone would

Figure S.1
The Transformational Objective of Alt#4

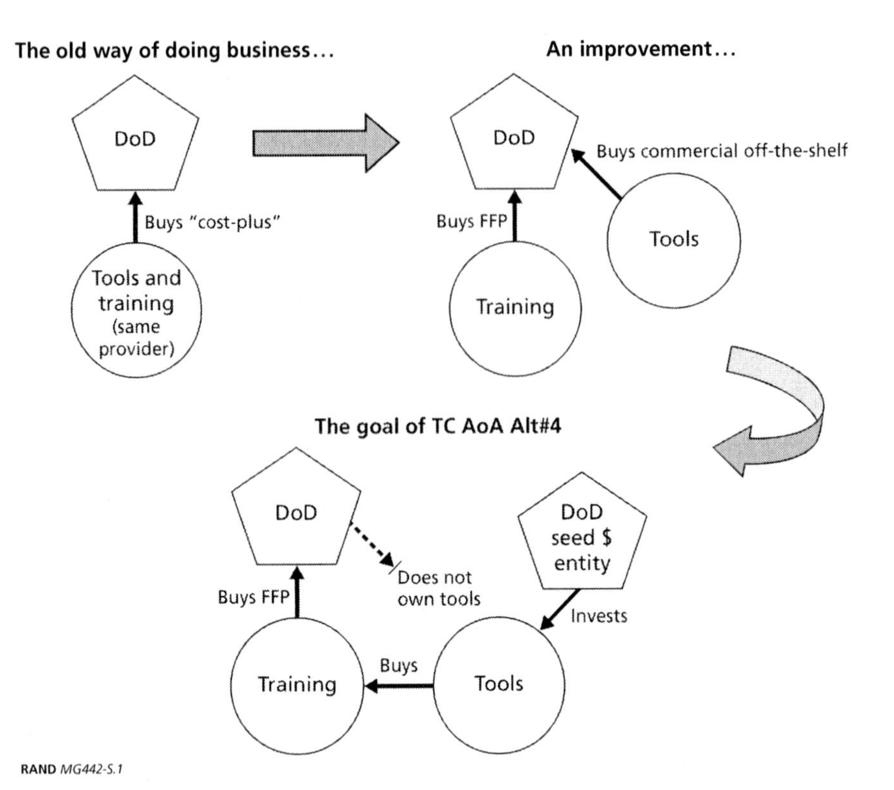

RAND *MG442-S.1*

significantly change the incentive structures for TSPs and tool vendors, but the designers of Alt#4 included additional elements in the business model aimed at obtaining further cost savings for the DoD and increasing DoD access to innovative simulation tools. Under Alt#4, the DoD would no longer buy simulation tools but would instead rely on TSPs to buy or license tools. To ensure a robust simulation tool market that meets the tool needs of the TSPs, the DoD would establish a seed money investment entity to stimulate the tool marketplace.

Other Business Models for Buying Simulations and Training

Chapter Three summarizes seven case studies: Joint Simulation System (JSIMS), Special Operation Forces Air Ground Interface Simulator (SAGIS), Virtual Convoy Combat Trainer (VCCT), Distributed Mission Operations (DMO), two examples from the United Kingdom Ministry of Defence (MOD) Naval Recruiting and Training Agency (NRTA)—Fire-Fighting Training Units (FFTUs) and Maritime Composite Training System (MCTS)—and the Central Intelligence Agency's (CIA's) In-Q-Tel venture capital corporation. Although the Alt#4 model has points in common with the other approaches, it has some distinctive features.

Table S.1 highlights several key differences between the different business models. The first critical difference is what the defense entity buys. The DoD (or MOD) buys simulation tools, simulation training, tools and training, or--in the case of DMO, FFTUs, and MCTS—availability of simulation tools. The core conceptual difference is between buying tools as goods (and then owning them) or buying tools as services (either directly or as part of a training service package).

Second, there are critical differences in the terms of the contracts under which the defense entity buys goods or services. Contracts are either cost-plus or firm-fixed-price and can be for either a short or a long duration.

The third critical difference is whether the simulation tools and the training come from the same contractor/provider. In several of the business models, delivery of training is wholly or partially integrated with provision of tools; in other models, tools and training are provided by different vendors.

Fourth, funding for tool development runs on a spectrum from public (funded by the DoD) to private (funded by contractors, tool makers, or the original equipment manufacturers). All the case study business models fall near one of the two extremes on this spectrum. Alt#4, however, offers the possibility of hybrid funding, where the

Table S.1
Features of Different Business Models for the Acquisition of Simulation Training

Model	Who Buys Tools	Who Funds Tool Development	Who Builds Tools	Who Owns Intellectual Property [*or Assembled Simulators*]	"Units" Tools Provided in	Who Provides Training	"Units" Training Provided in	Length of Contracts
JSIMS	DoD	DoD	Contractor	DoD (full government rights) [*DoD*]	Cost-plus	Same contractor	Billable contractor hours	Long and locked in
SAGIS	DoD	DoD	Contractor	DoD (full government rights) [*DoD*]	Cost-plus	Uniformed personnel	Classes	Acquisition length for tools, no contracts for training
VCCT	Contractor	Contractor	Contractor	Contractor [*Contractor*]	(a)	Same contractor	FFP person-hours of training	Relatively short
DMO	DoD	Contractor	Contractor	Contractor [*Contractor*]	FFP hours of simulator availability	Uniforms or different contractors	Varies	Long, performance extended tools contracts
FFTU	PFI contractor	OEMs	OEMs, subcontractor	OEMs [*PFI contractor*]	(a)	PFI contractor (with transferred MOD personnel)	FFP training days; excess capacity sold for MOD & PFI profit	Long
MCTS	PFI contractor	OEMs	OEMs, subcontractor	OEMs [*PFI contractor*]	Fixed-price for tool availability	RN uniforms and PFI contractor	FFP training days	Long
Alt#4	TSP contractor	OEMs, perhaps w/ catalog conductor seed money	OEMs	OEMs [*TSP contractor*]	(a)	TSP contractor	FFP per training outcome	Short

[a] Blank cell indicates that the training user, DoD or MOD, is not buying tools but is instead buying training outcome. The tool purchases are up to the contractor providing the training.

main responsibility for funding tool development lies in the private market place, but the DoD can contribute to tool development through seed money investments.

Economic Theory and Challenges Facing Alt#4

Chapter Four presents economic theory relevant to key aspects of Alt#4 and other observed business models. Theory relevant to the ownership of tools, cost-plus versus firm-fixed-price contracting, tied or untied markets, and the role of competition in innovation are all discussed. *The main conclusion we draw from the economics literature is that Alt#4 is based on sound economic principles and has good prospects for delivering efficiencies to the DoD's training community.* Our second conclusion is cautionary; theory suggests that the Alt#4 model is most likely to realize the cost efficiencies and innovations of the private sector when it is applied to technologies that also have commercial applications and training needs that are relatively straightforward to specify. Alt#4 may struggle if applied to certain technology/training areas.

Based on findings from economics, RAND experience with DoD acquisitions, and interviews conducted for this report, Chapter Five identifies six challenges facing the successful implementation of an Alt#4 prototype. These challenges and risks are not all of the same magnitude but each, if not dealt with effectively, could impede the success of the prototype. The discussion includes risk abatement strategies relevant to those challenges, where possible. The core risk/challenge areas are:

- setting operability/interoperability compliance standards that are neither too inclusive nor too exclusive;
- having a DoD component legally and effectively invest in a "venture-capital-like" fashion in the tool market;
- identifying and sharing new or emerging training needs;

- writing solicitations and contracts that allow providers to offer innovative best value solutions while ensuring that training needs are met;
- establishing effective performance measures for each firm-fixed-price contract let under the prototype; and
- resolving issues regarding risk, including:
  - transfer of cost uncertainty from the DoD to the TSPs;
  - risk to the DoD that providers may fail to manage their risks, go out of business, and not deliver contracted outputs; and
  - risk that prototype host/executor will implement a business model other than the desired full Alt#4 model.

Critical Elements of a Prototype Implementation Plan

Chapter Six lays out critical elements of a plan to implement a prototype of Alt#4. The RAND team framed a plan so that the prototype implemented following this guidance will

- be able to function legally within the DoD context;
- be true to the model as envisioned by the TC AoA business game team that conceived it;
- adhere to the model principles validated by economic theory in Chapter Four; and
- be well positioned to implement mitigation strategies against the risks identified in Chapter Five.

Full details are provided in Chapter Six. The implementation plan includes the establishment (or assignment to existing functions) of four core components: a governance/oversight entity; a tool catalog standards, sustainment, and investment entity we call the "catalog conductor"; an advisory board; and a contracting and grants office. Prototype implementation will leverage existing entities inside the DoD (one or more training users) and outside the DoD (the TSP and tool vendor markets). Chapter Six breaks down roles and responsibilities necessary for the prototype, and recommends allocations of those roles and responsibilities across prototype components.

This implementation plan recognizes that the prototype will have a limited budget ($15 million over three years). These funds need to support the operation of the core prototype components, including administrative costs and the personnel costs associated with executing the prototype. This budget is also the source for any seed money the catalog conductor invests in the tool vendor market and must cover costs associated with compliance testing as well. As much of the prototype budget as possible should be reserved for catalog conductor activities.

The prototype budget *should not* pay for training support. Training users already have funds with which to buy training and training support; the prototype simply asks them to do it in a new way. Training users should be either convinced to participate because of the efficiencies they are likely to realize through the prototype or coerced to participate through the DoD management structure or chain of command. Expensive incentives to training user participation, such as prototype budget funds for training support, should not be required.

Evaluating the Prototype

Chapter Seven presents an evaluation plan for the prototype. We recognize that implementing a program prototype takes a great deal of effort. Collecting evaluation measurements may not always receive the highest priority. Nevertheless, it is critical that the prototype executor collect and track sufficient data for a process evaluation, an outcomes evaluation, and an assessment of efficiency.

The best way to make sure that there are sufficient data for the evaluations is for the prototype executor to track data on critical transactions. Although evaluation is a core responsibility of the governance body, the action components (the catalog conductor, the contracting/grants support, and the training user) must carefully document their activities and expenditures. Even if critical data are just stored for future use, archived copies of all solicitations generated and all responses received (at all stages of the proposal and bid process) should be retained. When the catalog is established, catalog conductor personnel should make sure that the database in which it resides includes fields for date of certification and whether the vendor received seed money support from the catalog conductor or support from any DoD funding source (for example, the Small Business Innovation Research program).

In addition to keeping careful records on the DoD component side, effective prototype evaluation will require two active data collection efforts:

- satisfaction and performance surveys of the training users; and
- informational surveys of the training service providers.

Information about the TSPs' experiences—which catalog tools they considered, which they ultimately used, and their understanding of prototype policies—will make a significant contribution to evaluating the functioning of the prototype. Careful data tracking and plans for surveys of training users and TSPs should be added to the list of responsibilities presented in the implementation plan in Chapter Six.

Recommendations

We recommend that the DoD proceed with the prototype of Alt#4. The observed balance between theoretical plausibility and empirical risks suggests that this activity is highly appropriate to a test, pilot, or prototype project. We further recommend that the DoD strive to make the prototype implementation as close as possible to a test of the "pure" Alt#4 business model. Chapter Three shows that business models similar to Alt#4 can also be successful. But the Alt#4 business model contains some innovative elements not seen in existing approaches, and the prototype is an important opportunity to see those innovations in action.

For the prototype to succeed and for it to be an accurate representation of the Alt#4 business model, certain elements must be in place and several risks avoided or mitigated. Chapter Six contains RAND's proposed implementation plan for the Alt#4 prototype. The prototype also needs to include a training user who is willing to buy training under the rules of the prototype.

Finally, *we recommend that the DoD arrange for an impartial outsider to evaluate the prototype.* Evaluation by a nonstakeholder decreases the likelihood that parochial interest will play a part in the evaluation and increases the legitimacy of the evaluation.

Acknowledgments

Many people shared generously of their time and assistance as we pursued this research. We thank our sponsor representatives and project monitors in the offices of the Under Secretaries of Defense (OUSD) for Personnel and Readiness (P&R) and Acquisition, Technology and Logistics (AT&L), and the Joint Staff Operational Plans and Joint Forces Development Directorate (J7)—Fred Hartman, Skip Hawthorne, and Captain Jeff Miller—for their time, input, and support throughout the project. We wish to acknowledge all of the DoD personnel—uniformed, civilian, or contractor—who took time to share their relevant expertise regarding Alt#4, JCAS, simulation training, or contracting: Colonel Steve Howard and Mike Vaughn at SOCOM; Ken Goad, Warren Bizub, Marty Vozzo, Rob Grimes, Kathy Durbin, Colonel Dan Henkel, Lt. Colonel Andrew Riley, Edwin Weid, Gary Chiaverotti, and all the other helpful folks we met during or in correspondence after our repeated visits to JFCOM-JWFC in Suffolk, Va.; John Womack, Rick Mathews, and Lt. Colonel Russel Hinds from PEO-STRI in Orlando, Fla.; Bill Duncan from Live Fire Test and Training; U.S.Navy's John Bilbruck; Charles Colgrave and Bill Pattison of Air Combat Command, Langley, Va.; Defense Acquisition University's Dave Scibetta and John Krieger.

We would also like to thank Warren Katz (MaK Technologies), Bill Waite (AEgis Technologies), and the other members of the "Macrosystems" team at TC AoA business game #2 who developed the business model that became Alt#4 for sharing their understanding of the model with us.

We acknowledge and appreciate the time taken to speak with us by personnel in the U.K. training community, including the Royal Navy's Captain John Murphie, Flagship's Tim Redfern, and Commander Mark Foster and Jon Downing from the U.K. Ministry of Defence.

Thanks go to our RAND colleagues for their helpful comments during a work-in-progress internal seminar. Bruce Held, Cynthia Cook, Jeff Rothenberg, and Andrew Dick offered substantive follow-on help after this seminar.

We also thank our RAND colleagues Edward Keating and Matthew Lewis for their thoughtful and constructive reviews of this report as part of RAND internal quality assurance procedures.

Special thanks go to tireless RAND administrative assistants Maria Falvo and Samantha Merck for getting us where we needed to be when we needed to be there and for their critical assistance with document management, without which this report would be very much in shambles and sorely lacking complete references and citations.

Thanks to veteran RAND communication analyst Gordon Lee for helping us think about the structure of the report.

Errors and omissions remain the responsibility of the authors alone.

Acronyms

AT&L	Acquisition, Technology and Logistics
ATC	Air Traffic Control
AWACS	Airborne Warning and Control System
C4I	Command, Control, Communications, Computers, and Intelligence
C4ISR	Command, Control, Communications, Computers, Intelligence, Surveillance, and Reconnaissance
CAS	Close Air Support
CC	Catalog Conductor
CIA	Central Intelligence Agency
COI	Conflict of Interest
COTS	Commercial Off-the-Shelf
CPFF	Cost-Plus-Fixed-Fee
CPAF	Cost-Plus-Award-Fee
DARPA	Defense Advanced Research Projects Agency
DMO	Distributed Mission Operations
DoD	Department of Defense
FAR	Federal Acquisition Regulations
FFP	Firm-Fixed Price
FFTU	Fire-Fighting Training Unit

FY	Fiscal Year
GFE	Government Furnished Equipment
HLA	High Level Architecture
IC	Intelligence Community
J7	Joint Staff Operational Plans and Joint Forces Development Directorate
JCAS	Joint Close Air Support
JFCOM	Joint Forces Command
JNTC	Joint National Training Capability
JSIMS	Joint Simulation System
JWFC	Joint War Fighting Center
KPI	Key Performance Indicator
L, V, C	Live, Virtual, or Constructive
M&S	Modeling and Simulation
MCTS	Maritime Composite Training System
MOD	British Ministry of Defence
MTC	Mission Training Center
NRTA	Naval Recruiting and Training Agency
NTMS	Nontraditional Military Supplier
OEM	Original Equipment Manufacturer
OSD	Office of the Secretary of Defense
OT	Other Transactions
OTA	Other Transaction Agreement
OTC	Over the Counter
OUSD	Office of the Under Secretary of Defense
P&R	Personnel and Readiness
PEO STRI	Program Executive Office for Simulation, Training and Instrumentation
PFI	Private Finance Initiative
PIDS	Presidential Information Dissemination System

PPP	Public-Private Partnership
R&D	Research and Development
RFI	Request for Information
RN	Royal Navy
SAGIS	Special Operations Forces Air Ground Interface Simulator
SBIR	Small Business Innovation Research
SOCOM	Special Operations Command
TAC	Terminal Attack Control
TC AoA	Training Capabilities Analysis of Alternatives
TIA	Technology Investment Agreement
TSP	Training Service Provider
VCCT	Virtual Convoy Combat Trainer
VV&A	Verification, Validation, and Accreditation

CHAPTER ONE

Introduction: The Legacy of JSIMS

Launched in 1994, the Joint Simulation System (JSIMS) program was envisioned as an interoperable training simulation able to combine warfighting doctrine; command, control, communications, computers, intelligence, surveillance, and reconnaissance (C4ISR); and logistics for full spectrum joint warfare. It failed. After ten years and over a billion dollars in expenditures, JSIMS was canceled, with only a modest number of the systems it produced being in any way integrated into ongoing Department of Defense (DoD) training activities.

JSIMS is reviled not only as a failure in its own right but as an example of what is seen to be a larger set of problems with the DoD's core business approach to the procurement of simulations and simulation training support. In late 2002, while JSIMS was in its death throes, the Office of the Secretary of Defense (OSD) commissioned the Training Capabilities Analysis of Alternatives (TC AoA) to identify gaps in training capability and identify and assess alternatives for removing those gaps. The TC AoA was completed in July 2004. One alternative identified in the study, Alternative #4 (Alt#4 henceforth), proposed an innovative approach to the acquisition of simulation training support that would represent a fundamental change to the DoD's old way of doing business. The results of the TC AoA led

OSD to conduct a limited prototype of the Alt#4 business model over a three-year period beginning in FY 2006.[1]

This report accomplishes three tasks, all relevant to the proposed prototype effort. First, it considers the Alt#4 business model in light of existing economic theory and literature and relative to other business models for the acquisition of simulations and simulation training support. Second, it provides an implementation plan for a prototype of Alt#4, highlighting ways to best leverage the advantages of the model identified in task one and ways to minimize or avoid identified risks and pitfalls. Third and finally, it contains the elements of an evaluation plan for the prototype effort, so that at the conclusion of the prototype, stakeholders can tell whether it succeeded. An effective set of evaluations will allow stakeholders to determine whether the prototype was implemented as designed, whether it realized the goals and outcomes it was meant to, and its efficiency relative to previous comparable acquisition efforts.

The remainder of this chapter describes the problem, the TC AoA, and the Alt#4 model in greater detail. The chapter concludes with the details of RAND's tasking, a summary of findings, and the organization of the remainder of the report.

The Policy Problem

Although the cancellation of JSIMS inspired the TC AoA, the problem that Alt#4 hopes to address is broader and deals with the way the DoD has traditionally bought simulations and simulated training support. The old business model was fiscally wasteful and stifled innovation because it created a system of long-term commitments to contractor proprietary simulation systems that lack incentives for efficiency on the contractors' part.

---

[1] This is consonant with a view expressed by Held et al. (2002, p. xix), who advocate the DoD's use of prototypes and pilot programs: "This approach is consistent with the new industry paradigm that argues that one learns more about something by acting on it (in this case, by establishing pilot programs) instead of, as in the past, waiting until it is thoroughly understood before acting."

The exact features of the old business model that Alt#4 seeks to address are detailed in Chapter Two. This section briefly considers the broader problem and defines some of the terms that we use throughout the report.

At issue is how best to acquire simulations and simulation training support. When we say "best," we mean some process that offers the opportunity to combine reasonable cost, good value, effectiveness, and responsiveness to user needs.[2] The definitions below make clear exactly what we are talking about buying.[3]

Definitions

Here, *simulation* refers to a broad class of hardware or software that can be *live*, *virtual*, or *constructive* (L, V, C). DoD 5000.59-P defines these classes:

> Live, Virtual, and Constructive Simulation
>
> A broadly used taxonomy for classifying simulation types. The categorization of simulation into live, virtual, and constructive is problematic, because there is no clear division between these categories. The degree of human participation in the simulation is infinitely variable, as is the degree of equipment realism. This categorization of simulations also suffers by excluding a category for simulated people working real equipment (e.g., smart vehicles). Live Simulation: A simulation involving real people operating real systems. Virtual Simulation: A simulation involving

[2] A group of industry and DoD representatives meeting to discuss the innovative acquisition strategy concluded that the core goal of a business model in this area should be to "Develop and deliver timely, cost effective training using best products and practices, that is responsive to user needs" (JFCOM JWFC/JNTC Innovative Acquisition Strategy Offsite, August 3, 2005, BMH conference room, Suffolk, Virginia).

[3] We understand that "buy," "acquire," and "procure" can all take on specific technical meanings in the context of defense purchases. In common English usage, all three mean pretty much the same thing, and when we suggest that the DoD buy, acquire, or procure something as part of a business model under consideration, we intend to mean the same thing. When the purchase needs to be done in a certain way or with a certain contractual vehicle, we make that clear in the text.

> real people operating simulated systems. Virtual simulations inject human-in-the-loop in a central role by exercising motor control skills (e.g., flying an airplane), decision skills (e.g., committing fire control resources to action), or communication skills (e.g., as members of a C4I team). Constructive Model or Simulation: Models and simulations that involve simulated people operating simulated systems. Real people stimulate (make inputs) to such simulations, but are not involved in determining the outcomes.

The DoD vision of future training involves the combination of all three types of simulation into a single training experience or event; this is called *live, virtual, constructive integration*. Successful L, V, C integration remains challenging, but the potential benefits to military training are considerable. Imagine a fighter pilot flying a live simulated mission in an actual aircraft, able to see and interact with a wingman who is a pilot in a virtual flight simulator half a world away. Extend this formation into the constructive realm by adding a handful of computer-generated and computer-maintained pilots and aircraft, all of which can be perceived and interacted with by the two real pilots. Imagine a command post exercise where headquarters officers command and interact with forces training live at a training range, with additional forces participating virtually, with adjacent forces simulated constructively, and with an adversary force that contains live, virtual, and constructive elements.

In addition to being live, virtual, constructive, or a combination of the three, simulations also differ in their fidelity and complexity. Collectively, variation in these dimensions is summarized when a simulation is referred to as lightweight or heavyweight. *Heavyweight* simulations push toward the highest levels of fidelity and realism and often require special hardware, either sophisticated interface equipment or computers more powerful than standard PCs. The classic example of a heavyweight simulation is the fully enclosed cockpit trainers used for rotary or fixed-wing aircraft, which often have full canopy enclosures and simulate extremely realistic physics/physical conditions. The pilot faces "real" dashboard instrumentation, and the canopy is fully enclosed in a virtual "sky." These might also include a hydraulic system to change the angle of the simulator's cockpit and provide movement and directional sensations. The consequences of speed, altitude, angle, and wind are modeled for the simulated aircraft in very high detail (high fidelity). See Figure 1.1 for a classic example of a heavyweight simulation.

Lightweight simulations are lower fidelity and more easily accessible. The fidelity of lightweight simulations is accurate enough for the given purpose of the simulation. Lightweight simulations usually run on PCs, game consoles, or other off-the-shelf hardware, and the simulation is often sufficiently straightforward and simple for the trainee to use without the assistance of an operator. Examples of lightweight simulations include PC-based flight simulators, PC-based

Figure 1.1
The MH-6M Light Assault/Attack Reconfigurable Combat Mission Simulator

SOURCE: Hinds (2005).
RAND *MG442-1.1*

strategy "game-"driven training, and PC- or gameconsole-based "first person" simulations such as the "America's Army" game/recruitment tool/training tool.[4]

Although the end user's simulated experience may be lightweight or heavyweight and may be live, virtual, constructive, or a combination thereof, the simulation itself may be assembled from a host of modular simulation tools. *Simulation tools* are the hardware or software components that go into creating the overall simulation. A simulation might include software for terrain modeling, a model for weapons effects, a set of algorithms to determine object trajectories, viewable three-dimensional objects to represent vehicles and persons, algorithms that determine the behavior of constructive noncombatants, as well as the hardware necessary to run the simulation (one or more PCs or more powerful computers), interact with the simulation (steering wheels, joy sticks, light guns, and sensors to detect the emitted light), and perceive the simulation (computer monitors, projection screens, and dashboard gauges). In addition to the software tools that combine to generate and maintain a simulated environment and the hardware tools that display that environment and allow the users to interact with it, a variety of integration and networking tools can help the software connect with the hardware or other software, or enable multiple virtual participants to network into the same simulated environment, or convert imagery from real places into virtual terrain and building models (as just a few examples). Finally, there are a host of training support software elements, such as scenario authoring and preparation software, exercise management software, data logging tools, and after action report generation and analysis tools. All of these can be considered simulation tools.

Training support refers to the contractor support necessary to train with simulation tools. Training support could range from a minimal amount of technical assistance (installation and maintenance) to the complete outsourcing of training for a particular event or skill.

---

[4] This can be downloaded free from http://www.americasarmy.com/. It requires a reasonably powerful graphics card to run.

Acquiring Simulations and Training Support

DoD use of simulations for training is likely to continue to increase for several reasons. First, simulations allow the DoD to save money in a number of ways: by reducing wear on vehicles, minimizing fuel and ammunition costs associated with traditional live training, and avoiding costs associated with moving personnel and materiel to training sites. Second, ever-improving simulation tools allow forces to effectively train on a wider range of tasks through simulation training. Third, simulations allow training for many tasks that are dangerous, costly, or environmentally or politically unsuitable for live training. Such tasks might involve high explosive ordnance; force-on-force training using the full spectrum of available weapons; operating under nuclear, biological, or chemical attack conditions; exercising military formations dispersed over a wide area of operations; or operating aircraft in questionable weather conditions.

We expect the use of simulated training to continue to increase, and we recognize that training remains one of the principal occupations and obligations of U.S. military personnel. We also recognize that many simulation tools require operators with complex technical skills that are not traditionally part of military core competencies. Where the DoD uses simulations for training, use of some training support from sources external to the training unit seems likely.

Which tasks U.S. forces should train live, virtual, or constructive is a policy decision beyond the scope of this report; so is the decision to fully outsource or proceed with more limited training support for tasks that will be trained through simulation. Regardless, the DoD will be buying simulation training support and should seek to do so in a cost-effective, best value fashion.

The Training Capabilities Analysis of Alternatives and the Alt#4 Solution

Through the TC AoA, the DoD sought to generate and analyze alternatives to the old way of buying simulation training. The TC AoA process involved study and analysis by a core team, supported by input from three expert panels: a training panel, a technology panel, and a cost panel.[5] The process also included several "business games" during which persons from industry, academia, and government collaborated to focus the analysis and inject new ideas. One business game (business game #2) grouped industry participants into teams and encouraged them to propose a business model for the procurement of simulation training. One team (called the "Macrosystems" team during the game) developed an innovative acquisition strategy, which, when fleshed out, joined the alternatives being considered in the TC AoA as Alt#4 (Garrabrants et al., 2004).

The Transformation Envisioned Through Alt#4

The logic underlying each element of Alt#4 is presented in the next chapter. Briefly, the model proposes that the DoD change the way it buys training support and simulations so that the financial incentives of the training support providers (TSPs) and the tool vendors align toward innovation, delivery of quality training, and cost savings for the DoD, the TSPs, and the tool vendors. As one of the TC AoA outbriefs to the DoD senior steering group put it, the goal is to "buy training and let market demand develop enhanced capabilities."

Figure 1.2 displays the transformation envisioned by Alt#4. The designers of Alt#4 maintain that the biggest problems with the old way the DoD bought simulation training stem from the vertical integration of tool and training provision in single contractors and the use of "cost-plus" contracts for the procurement of training support and simulation tools. Under the Alt#4 business model, the DoD would use firm-fixed-price (FFP) contracts instead of cost-plus contracts, and the tool market would be separated (untied) from the TSP market. Although those two changes alone would significantly alter

[5] See Office of the Secretary of Defense and United States Joint Forces Command (2004) for further details.

Figure 1.2
The Goal of Alt#4

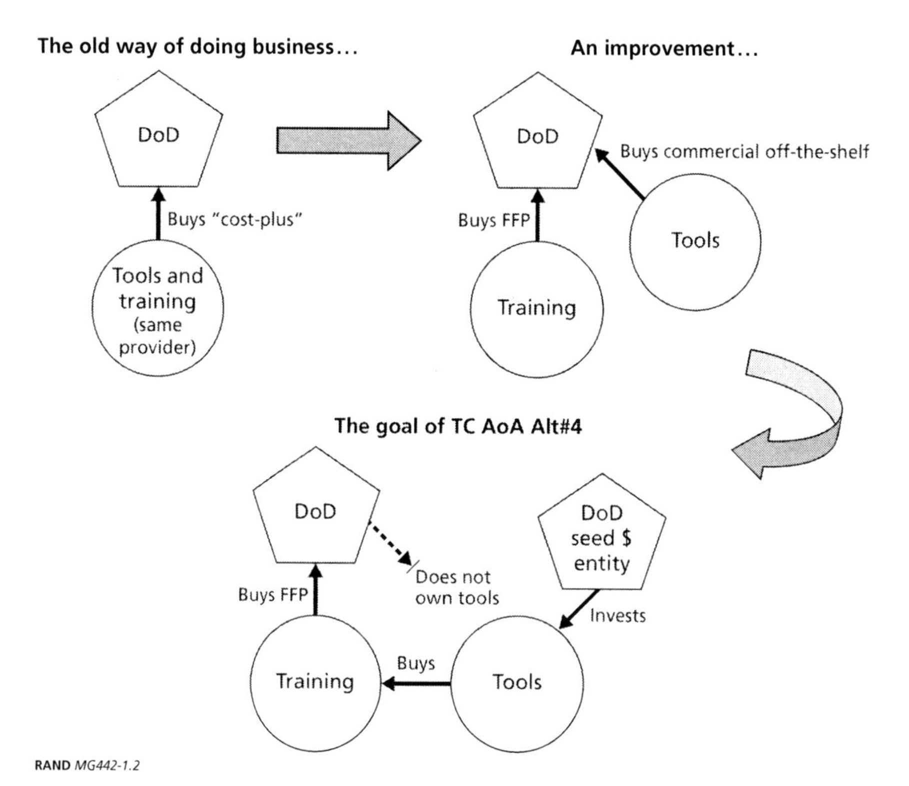

RAND *MG442-1.2*

the incentive structures for TSPs and tool vendors, the designers of Alt#4 included additional elements in the business model aimed at further cost savings for the DoD and increasing DoD access to innovative simulation tools. Under Alt#4, the DoD would no longer buy simulation tools, instead relying on TSPs to buy or license the training tools they require to deliver the training. To ensure a robust simulation tool market that meets the tool needs of the TSPs, the DoD would establish a seed money investment entity to stimulate the tool marketplace. Full details can be found in Chapter Two.

RAND's Tasking

Our sponsors in the Offices of the Under Secretaries of Defense (OUSD) for Personnel and Readiness (P&R) and Acquisition, Technology and Logistics (AT&L), and the Joint Staff Operational Plans and Joint Forces Development Directorate (J7) asked RAND to accomplish two core planning and analysis efforts for the prototype: an implementation plan and an evaluation plan. Subtasks in service of these goals included an evaluation of the merits of the Alt#4 business model itself, based on economic theory and business experience, and the collection of observations and lessons learned from existing efforts to buy simulations or training, both domestically and internationally. This report presents findings from these efforts.

Findings in Brief

Our findings are based on analyses from two data sources: first, a review of documents, literature, and theory from economics and publicly available information about various simulation training initiatives; second, interviews with industry and DoD participants in the TC AoA process or with experience developing, procuring, or using simulations for training. We also interviewed personnel from both the industry and defense side of simulation training provision in the United Kingdom. From these data, we assembled business model case summaries of relevant programs and a set of business models that correspond to approaches to buying simulation tools or training support.

This analysis finds that the economic principles underlying Alt#4 business model are sound. The paragraphs below list our key findings.

- From the perspective of economic theory, the incentive structures created by an implementation of the Alt#4 business model can encourage training service providers and tool vendors to provide tools and training in a cost-efficient manner; in the abstract, Alt#4 could work as designed.

- However, DoD procurement experience and economic theory suggest that there are risks associated with efforts to implement several of the core elements of the Alt#4 business model. Failure to successfully mitigate these risks could result in the unintended creation of perverse incentive structures, or other failure to realize the full potential of the model, or the failure of the model entirely. Challenges include
  - Setting operability and interoperability standards for tools that are neither too exclusive nor too permissive.
  - Writing training solicitations and contract performance measures that are effective for FFP contracts.
  - Applying the Alt#4 business model in areas of training and simulation technology where economic theory suggests it is most appropriate.

These findings lead us to recommend the following:

- The DoD should proceed with the prototype. The actual implementation of the Alt#4 business model will reveal which risks are genuine threats, which mitigation strategies are effective, and the extent to which Alt#4 actually encourages cost efficiency and innovation in the simulation tool market.
- Implement the full Alt#4 model, not a hybrid or reduced model. Although the analyses in Chapters Three and Four suggest that a single core element of Alt#4—switching from cost-plus to firm-fixed-price contracts—provides a significant part of the benefits possible, the prototype is a great opportunity to see what else the DoD stands to gain through the implementation of the other innovative aspects of the business model.
- Adopt the risk-mitigation strategies outlined in Chapter Five. Chances of prototype success increase with each implementation challenge that is successfully resolved.

Organization of the Report

Chapter Two presents the theoretical underpinnings of the Alt#4 business model. It describes the elements of the old business model that Alt#4 proposes to improve on and the details of the solution. Chapter Three collects observations from case summaries of other efforts to procure simulation training. Chapter Four considers the Alt#4 business model (and some of the business model elements from the other cases) in light of theory and research in economics. Drawing on the results of Chapters Three and Four, Chapter Five highlights the challenges that will face the prototype effort. Chapter Six presents the critical elements of an implementation plan for the prototype, including the roles and responsibilities of different components of the prototype executor and a recommended order in which to begin prototype activities. Chapter Seven details an evaluation plan for the prototype. Chapter Eight concludes with a discussion of challenges that will face stakeholders in their efforts to broaden the application of the Alt#4 business model beyond the prototype if the outcome of the prototype is satisfactory.

CHAPTER TWO

The Alt#4 Solution

The solution that Alt#4 proposes to DoD's simulation training procurement woes has two core elements: First, the DoD stops buying both tools *and* training support and instead buys only training support; second, the DoD stops buying training support with cost-plus contracts and starts buying it on FFP per training outcome contracts.[1] Alt#4 proposes additional DoD efforts to ensure innovation and competition in the simulation tool market:

1. separating the training support and tool markets through conflict-of-interest clauses in training support contracts (referred to as "untying the markets" in the economics literature);
2. imposing compliance with adopted technical standards to ensure product operability and interoperability; and
3. creating a mechanism for the injection of seed money into the tool market to support innovation or create competition in needed submarkets.

The logic underlying each element is to have the DoD eliminate the perverse incentive structures of previous business models and in-

[1] Alt#4 as envisioned by its creators would replace existing models for the acquisition of simulations and simulation training support DoD-wide. We realize that Alt#4 is presently being considered only for a limited prototype and that the decision to buy tools and training as a single service may not be appropriate to all simulation and training efforts, but we present the model in this chapter in the broad way in which it was conceived.

stead create incentives for the training service providers and tool vendors to provide responsive simulation training support using best products and practices at best value prices.

We begin this chapter with a discussion of the problems with the DoD's old simulation training procurement business model addressed by Alt#4. Then, we detail each core element included above and conclude the chapter with a brief summary of the Alt#4 model, including some of the processes and interactions implicit in the model elements.

Problems with the Old Business Model That Alt#4 Seeks to Address

The underlying logic of the Alt#4 business model involves the DoD making sure that all participants' economic incentives align toward efficiency, that is, delivering good training with good tools at optimal prices. Alt#4 presupposes two core problems with the old way the DoD acquired simulators and training support, each bringing with it a small handful of specific misaligned incentives. The two core problems are owning the simulation tools and using cost-plus contracts to buy simulations and training support from the same provider.

DoD Ownership of Tools

Under the old business model, the DoD would procure simulations in much the same way it procured major weapon systems. For simulations, this would involve a contractor producing the simulation tools (hardware and software) and delivering them to the DoD, which would then own the tools, usually with full government rights (as opposed to the contractor retaining proprietary rights). DoD ownership results in several inefficient incentive structures:

1. Because the DoD owns the tools, there is a strong incentive to keep using them (they are already paid for), even if new or different tools would be better or more appropriate for new training

needs (also referred to in the economics literature as a high switching cost).

2. Because the tools are DoD-owned, the DoD must pay for any desired upgrades or improvements to the tools. Because of the high cost associated with a switch to new tools, old tools are kept in service but new money is spent trying to rejuvenate or update them. Tool vendors providing these upgrades have no incentive to efficiency either because of their cost-plus contracts (see below) or because they are the sole possible source for the upgrades and face no competition.
3. Even though the DoD owns these tools, the knowledge necessary to operate the simulations (the training support) either is outside the core competencies of the DoD or is sufficiently arcane to be de facto proprietary to the contractor that made the tool, so the DoD pays even more money each time it wants to use the tools it owns.[2]
4. Although vendors compete to provide the simulation tools to the DoD in the first place (ex ante competition), once the contract is signed, there is no competition for the subsequent support for and use of those simulation tools (no ex post competition).

Cost-Plus Contracts

Because of the potentially changing needs of the military, the importance of meeting military needs, and the uncertainty surrounding costs associated with developing new technologies and tools to meet military needs, many DoD contracts are written as cost-plus contracts. Cost-plus is short for cost-plus-fixed-fee or cost-plus-award [or incentive]-fee—an arrangement whereby the contractor is reimbursed for the real costs it incurred, *plus* a fixed profit or profit margin based on expected costs with some possible variation depending on incen-

---

[2] This is the so-called "men in white suits" problem. One of our interview respondents evocatively described the problem and indicated his frustration with the DoD paying the tool vendor hundreds of thousands of dollars to "send a hundred men in white suits" to run the DoD-owned simulations every time they wanted to put on a training exercise.

tive clauses.[3] Contracting in this way allows the vendor to shift cost uncertainty risks to the DoD, and, in return, the DoD has confidence that the vendor will not go out of business but will deliver the contracted goods or services, whatever the cost. Strict (and costly) government auditing is required to make sure that "costs" are appropriately accounted for. (See a further discussion of the pros and cons of cost-plus contracting in Chapter Four.)

The DoD's old business model for simulation procurement used cost-plus contracts. Alt#4 argues that two perverse incentive structures stem from cost-plus procurement of simulation tools and support:

1. Cost-plus contracts lack general incentives to efficiency. Costs are reimbursed fully, and profit is a proportion of expected costs. Because the DoD reimburses the contractor for all of its allowable costs, there is little incentive for the contractor to seek cost savings.[4]
2. In the same vein is the incentive structure for a cost-plus contractor faced with a make or buy decision. Imagine a new simulation array that requires a new widget to function. The cost-plus vendor must choose to either buy the widget from an existing commercial off-the-shelf (COTS) vendor at a modest price or make the widget at a much higher price (effectively reinventing the wheel, or widget in this case). The cost-plus contract rewards the contractor for choosing the more expensive option: The contractor can employ more personnel and pay them to engage in widget creation and have the cost paid without an adverse impact on profits.

---

[3] FAR 15.404-4(a)(3) and 15.404-4(c)(4) limit these statutory fees to 15 percent of the estimated contract cost for experimental, developmental, or research work and 10 percent for all other cost-plus contracts.

[4] Also note that in a cost-plus contract, there is no incentive for a contractor to tell the DoD that specified requirements cannot be met. We have heard anecdotally that contractors are more than willing to overpromise and then spend heavily at "cost" in pursuit of an unattainable goal. JSIMS might even be an example of this.

Key Elements of the Alt#4 Business Model

Alt#4 seeks to align the economic incentives of all participants toward efficient delivery of simulation training support and to eliminate perverse incentives identified in the old business model. Alt#4 proposes two core changes in business approach to support that goal and a further tripartite effort to ensure innovation and competition in the tool vendor marketplace. The two core changes are:

1. The DoD stops buying simulation tools and training support and buys only training support, with the TSPs buying or licensing the tools directly from the tool vendors or original equipment manufacturers (OEMs).
2. The DoD buys training support with FFP per training outcome contracts instead of cost-plus contracts.

The three elements of the effort to ensure innovation and competition in the tool vendor marketplace are:

1. untying the TSP market from the tool vendor market through enforcement of restrictive conflict of interest clauses in the FFP contracts written with TSPs;
2. establishing operability and interoperability standards for tools; and
3. injecting seed money into the tool vendor market to encourage innovation and create (or threaten to create) competition.

Buying Only Training Support

As noted above, DoD ownership of simulation tools leads to a handful of inefficient incentive structures. The Alt#4 model proposes to solve this problem by getting the DoD out of the business of owning simulators for which it will also need training support.[5] If the DoD

[5] The range of training events and types of simulators that Alt#4 is appropriate to remains an open question. Advocates among the industry business game participants who invented Alt#4 optimistically assert that it is applicable to all kinds of simulation training. An internal

contracts for training support only and leaves the choice of (and the procurement of) the tools to the training service providers, the DoD rids itself of perverse incentives to keep using and investing in aging simulations that have been overtaken by innovations in modeling and simulation technology. Additionally, the so-called "men in white suits" problem, where the DoD is forced to pay tool vendors to come and operate the tools the DoD "owns," disappears. Under Alt#4, the men in white suits bring their own tools, and if the DoD is not satisfied with the men, the tools, or the prices they charge, the DoD can re-compete the contract and hire a different TSP without facing an excessive "switching cost." This leads to the final proposed improvement in incentives: Because the DoD does not own the tools, training support can be bought when the contract is first bid and also when the contract comes up for renewal or re-competition. (New competitors will not be excluded from bidding because they lack the de facto proprietary knowledge the original tool vendor has of the "DoD-owned" tools, as happens under the old business model.)

Buying Training Support with Firm-Fixed-Price Contracts

If the DoD buys training support on firm-fixed-price contracts, several perverse incentives on the part of the contractors disappear. First, contractors have incentives to save money; the less it costs them to deliver the training, the more profit they make. These incentives include making the "right" decision in a "make or buy" situation; reinventing the widget will be much less likely, unless it happens to be the less expensive option (this, of course may depend on the health and competitiveness of the tool market, which is part of the motive for Alt#4's additional measures).

RAND audience expressed some concern as to whether the commercial market would be willing to provide military-specific or heavyweight simulators without contracting directly with the government. Although this remains a concern, the success of distributed mission operations (DMO) (discussed in Chapter Three) in providing fixed-wing aircraft cockpit simulation training (military-specific and heavyweight) without the DoD owning the simulators suggests that Alt#4 may be more broadly applicable than critics assume. The importance of both broadly adopting a more efficient business model across the range of goods and services it is applicable to and maintaining flexibility to meet the training needs of warfighter in the best way receives further attention in the concluding discussion in Chapter Eight.

Ensuring Competition and Promoting Innovation in the Simulation Tool Market

For the training service providers to be able to meet the DoD's solicitations for training support, simulation tools must be available to the providers. For the best tools to be available to the TSPs at optimal prices, a robust tool vendor market must exist that is both innovative and competitive. Alt#4 proposes several steps to ensure a robust, innovative, and competitive tool vendor market.

If TSPs are allowed to be tool vendors as well, they would have incentives to use their own cheaper, "good enough" tools rather than buy better tools from other tool vendors. To prevent this problem and to encourage competition both in the TSP market and in the tool vendor market, Alt#4 requires that the two markets be separated by strict conflict-of-interest clauses. In the firm-fixed-price per training outcome solicitations that the TSPs bid on, clauses are needed to restrict them from using tools of their own design. Vendors that traditionally are both TSPs and tool vendors must decide which market to participate in, or else they must spin off one of their units to meet conflict of interest requirements. This untying will allow simulation tools to compete fairly for use by the TSPs (giving them incentives to use best value tools) and will allow TSPs to compete fairly for DoD training support contracts (without one TSP having an advantage in the form of a proprietary tool).

Alt#4 also proposes to establish a catalog of certified simulation tools that the TSPs will purchase or license for the provision of training support. Tools listed in the catalog must pass compliance testing against standards for operability and interoperability. The DoD will establish these standards (or adopt existing standards) and arrange for compliance testing. TSPs "shopping" for tools in the catalog will know that listed tools work and are certified to be "plug and play" with the other tools in the catalog. Under Alt#4, the DoD contracts with TSPs must specify that they may use only catalog-certified tools.

Standards of this kind are critical to ensure competition in the tool vendor market. If tool vendors and OEMs are allowed to sell tools that will not integrate with the tools of others to TSPs, the modular independence of the simulation tools breaks down. No longer would a TSP be able to assemble a host of modular simulation tools into a best value whole by choosing the best value at each modular juncture; if the TSP wished to include a tool that functions only on a proprietary protocol, it would be stuck with the other modules conforming to that protocol, even if they are not the best value.

Finally, Alt#4 recognizes that the tool vendor market may need some external assistance to innovate in directions that will serve emerging DoD training needs and support multiple competing products in critical product areas. Alt#4 proposes that the DoD strategically inject seed money in a "venture-capital-like" way.[6] These seed investments could either be to encourage the development of new tools that will meet existing or emerging unmet training needs or to create competition (or the threat of competition) with tools unique in the market to discourage the vendors of those tools from engaging in monopoly pricing.

The focus of these seed money investments is on making sure that needed tools enter the catalog and that TSPs do not face monopoly pricing. Seed money investments should not be necessary where the existing market is sufficiently robust or where new tools have extensive commercial application. If tool vendors stand to profit independent of the DoD's needs, the addition of the DoD's needs to potential profits should be more than sufficient to foster innovation following standard commercial practices. Only when tools have limited potential commercial application, or potential commercial yields are too low to inspire independent investment in new technologies, should DoD seed money investment be necessary.

---

[6] "Venture-capital-like" because no equity stake is required; indeed, the DoD could face serious legal repercussions if it were to end up owning an equity stake in a private sector firm. This venture-capital-like seed money could either be a contract vehicle or a disbursement that allows the DoD to seed technological development, or it could be actual venture-capital investment conducted through a nonprofit venture-capital corporation established for the purpose, such as the Central Intelligence Agency's (CIA's) In-Q-Tel. See the discussion in Chapter Five for further details.

Summary of Alt#4

The discussion above presents the core elements of the Alt#4 business model and the motives for those elements. The full Alt#4 model as developed out of the TC AoA also includes several practical suggestions regarding realizing the business model; these are addressed in Chapter Six, which presents RAND's implementation recommendations. This section summarizes the processes and interactions implicit in the core elements of Alt#4 with a slightly less abstract example of a single pass through the business model.

Because the business model is a continuous process, it is difficult to choose a point from which to start an example. With that caveat, imagine a situation in which a new training need emerges, and it appears to be a need that can be met with simulation-based training, which can either be wholly outsourced or can take advantage of significant training support from contractors.[7] Once the need is identified, it must be shared throughout the DoD and the TSP market (so that they can begin thinking about satisfying the need) and the tool vendor market (so that tool vendors can begin thinking about new tools or modifications to existing tools to satisfy the need). The mechanisms for this dissemination of information are implementation-dependent, but the DoD might announce an emergent training need in a host of ways. Shortly after the need is identified, the DoD entity in charge of strategic seed investments in the tool market should assess the ability of existing tools to meet the need and the competitiveness of the tool market segments the need will rely on. As Chapter Six describes, this daunting task is facilitated by an advisory board with members from industry, academia, and government. With assistance from the advisory board and with information from its own market research, the seed money component decides whether new products will be required, either to meet the need or to ensure competition to meet the need. In either case, the seed money compo-

[7] For example, when insurgents and improvised explosive devices began to threaten U.S. convoys in Iraq, a need to better prepare forces for convoy operations emerged. Virtual Convoy Combat Trainer (VCCT), described in Chapter Three, was developed to meet this need.

nent then prepares a solicitation indicating what the need is, what kinds of simulation tools might be appropriate to meeting the need, and what kind of seed money support the DoD is prepared to offer. The seed money component will receive proposals and make awards to firms to help fund new products.

Simultaneous with this activity, training users will prepare statements regarding their new training need and will think about ways to demonstrate or certify satisfaction of this new need. Training users will prepare a solicitation for training support that does not specify exactly how the training must be delivered. Rather, it will present the training need as a list of training goals or "training targets" with as few fixed requirements as possible. This gives the TSPs the opportunity to bid innovative solutions to meeting the need and the flexibility to bid cheaper "90 percent solutions" that the training user can then evaluate for best value. Note that the TSPs can also determine how to certify satisfaction of the training need, as it is in their interest to have the customer satisfied with the training received.

In preparing their bids in response to the training users' solicitation, TSPs must also consider the tools that will be required to deliver the training. They can do market research, considering tools already certified into the tool catalog, past performance of the tool market, and the solicitations made by the seed money component. Working from their prior experience, they can assess risk associated with uncertainty about the cost or availability of required new tools and base the "safety margin" of their bid accordingly.

Depending on the urgency of the training need and the speed of the tool vendor's development processes, new tools may emerge before or after TSPs have submitted their bids. When they are complete, new tools are submitted for compliance testing and, when found in compliance with pre-established standards (which can be periodically reviewed and updated as necessary), are entered into the certified tool catalog.

Training users receive FFP bids from TSPs in response to their solicitation. Training users select the best value bid and engage in final contract negotiations to make sure the training certification procedures agreed on are effective measures of training success. This FFP contract also contains the standard Alt#4 business model conflict of interest clause requiring that TSPs not be tool vendors and that tools used in the training delivery comply with established standards.

The TSP that wins the contract arranges to buy or license tools from the catalog (or those about to enter the catalog) and prepares to deliver the training support.

DoD oversight monitors the whole process to make sure that the model is working as planned, that laws and regulations are adhered to, and that DoD components struggling with the challenging aspects of the business model have the guidance and support they need.

Of particular note are what does *not* occur during this process:

- The DoD does *not* buy or come to own any tools.
- DoD auditors do *not* descend on the TSP to inspect its books; if the training is satisfactorily delivered, then the TSP is paid the agreed-on fixed amount; its costs are its costs and of no concern to others.
- Tool vendors do *not* experience any of the frustrations of having the DoD as a customer. Unless they are taking seed money from the DoD, they have no interaction with the DoD at all, beyond submitting their tool to compliance testing, the functional application of which has likely been outsourced by the DoD.
- No one gets paid on cost-plus contracts. Tool vendors are competing to deliver tools to the TSPs, and the TSPs are competing to deliver training support to the DoD. Everyone has an incentive to be competitive with all the positive connotations of the term.

Table 2.1 summarizes the differences between the Alt#4 business model and the old model of doing business. Alt#4 differs from the traditional business model in all eight of the aspects detailed in Table 2.1: who buys the tools, who funds the development of the tools, who builds the tools, who owns both the intellectual property of the tools and the assembled simulations, the units the training

Table 2.1
Differences Between Alt#4 and the Old Business Model

	Old Model	Alt#4
Who buys tools	DoD	TSP contractor
Who funds tool development	DoD	OEMs, perhaps with DoD seed money
Who builds tools	Contractor	OEMs
Who owns intellectual property	DoD (full government rights)	OEMs
Who owns assembled simulators	DoD	TSP contractor
What units are tools provided in	Cost-plus	As agreed between TSP and OEM
Who provides training	Same contractor	TSP contractor
What units is training provided in	Billable contractor hours	FFP per training outcome
Length of contracts	Long and locked in	Short

is provided in, who is providing the training, the units the tools are provided in, and the length of the contracts.

In the next chapter, we consider other demonstrated approaches to the procurement of simulation training and compare them with the old business model and the business model proposed by Alt#4.

CHAPTER THREE

Other Approaches to Buying Simulation Training: Case Examples

This chapter considers other ways to buy training support or simulations by looking at the business models implicit in other acquisitions by the DoD, the CIA, and the British Ministry of Defence (MOD). Observations and lessons to be learned from these case summaries are both positive (examples of success) and negative (examples of less than optimal outcomes). These cases show, in the aggregate, that there are several different potentially successful approaches to acquiring simulations and training support. These different approaches rely on sometimes quite different economic logics. Taken together, these cases suggest that several different business models may be appropriate for different parts of the simulation and simulation training support markets.

There Are Many Ways to Buy Training and Simulations

This chapter presents case summaries for seven case examples: JSIMS, Special Operations Forces Air Ground Interface Simulator (SAGIS), VCCT, DMO, two examples from the British MOD (Naval Recruiting and Training Agency [NRTA]—Fire-Fighting Training Units (FFTU) and Maritime Composite Training System [MCTS]), and the CIA's In-Q-Tel venture-capital corporation. The chapter concludes with a summary and comparison of the various business models.

Selected Case Examples

For each case, we summarize some background details regarding the program, the systems involved and the uses they are put to, the outlines of the implicit business model (as far as they can be discerned for each case), and observations or lessons learned that are relevant to the current inquiry.

Joint Simulation Systems

Launched in 1994, the JSIMS program was designed to provide interoperable training simulations able to combine warfighting doctrine, C4ISR, and logistics for full spectrum joint warfare. JSIMS was expected to yield significant cost savings compared with other programs associated with integrating disparately conceived, designed, and documented legacy systems (Griffin et al., 1997). One stated objective of the JSIMS program was to cut the services' simulation operation and maintenance costs by two-thirds. JSIMS development began in 1996.

JSIMS sought to provide a central and unifying "architecture" for simulations across the service branches. In the JSIMS concept, a common simulation engine included system software that would enable JSIMS to run on computer hardware and networks that were already commercially available. JSIMS exercises would be run in a distributed fashion by relying on the then newly established DoD High Level Architecture (HLA) standards for simulation compatibility. Ideally, the HLA would create "interoperability among simulations and promote reuse of their components" (Defense Modeling and Simulation Office, 1995).[1]

---

[1] Note that Davis and Anderson (2003) find that HLA standards have effectively fostered simulation interoperability but suggest that it might be an opportune time to revisit and update them.

Before JSIMS, each service maintained a set of simulations within its own "stovepipe" that were not cross-compatible with each other. The Army's Core Battle Simulation system used a hexagon-based grid to simulate terrain, whereas the Navy's Research and Evaluation Simulation Analysis system relied on latitude and longitude measurements, and the Air Force's Air Warfare System offered a third representation, very different from those of the other services (Slabodkin, 1997). Had it worked, JSIMS would have resolved this incompatibility.

The many technical/tactical requirements and differences between the services could not be resolved without spending increasingly prohibitive sums. JSIMS has been called a "billion dollar boondoggle," because of the total amount of money spent on JSIMS itself and the service projects to which it was connected (Strategypage.com, 2005).

In December 2002, OSD issued a directive canceling JSIMS because of delays in development and cost overruns. The remaining JSIMS applications and aims were turned over to the Joint War Fighting Center (JWFC) for future development (Tiron, 2003).

JSIMS Business Model

JSIMS was developed with cost-plus contracts—cost-plus-award-fee (CPAF) and cost-plus-fixed-fee (CPFF) agreements.[2] Had JSIMS ever been completed, the reasonable surmise is that training support for JSIMS systems would have come from the same contractors on an ongoing cost-plus basis. JSIMS components that were completed are owned by the DoD with full government rights, meaning that the DoD owns the intellectual property foundation of the simulations, not just the purchased instances of the tools.[3]

---

[2] CPAF contracts provide an estimated cost, plus a fee of a base amount and an award amount. The government makes periodic evaluations of the performance of the contractors, the results of which determine the award fee amount. The government reimburses all allowable costs in addition to a fixed dollar amount in CPFF contracts.

[3] This is an important distinction in software ownership. Most of the time, software is just licensed to purchasers and the software vendor retains ownership. For example, all of the

Each service supported the larger program by funding its own next-generation models: the Army's War Simulation-2000 (now called WARSIM), the Air Force's National Air and Space Model, and the Navy's Maritime System. These systems were also acquired with cost-plus contracts. The JSIMS business model is representative of the old way of doing business.

Although JSIMS never delivered what it had been intended to, its remnants survive and have been incorporated into some ongoing DoD training. Each service made its own section of JSIMS usable, with varying degrees of success (Strategypage.com, 2005). The program was considered to be a sufficiently significant failure to spark the TC AoA.

Lessons Learned from JSIMS

JSIMS was ultimately too ambitious and lacked the incentive structure necessary to get all the participants (both the contractors and the services) moving with best efficiency in the same direction.

Root, Osterheld, and McAuliffe (n.d.) quote CDR Brian Hudson's summary of the JSIMS failure:

> JSIMS failed due to the separation over time between the real training objectives (the goals) and the actual implementation. The JSIMS effort got side tracked in trying to manage "cost vs. fidelity" instead of "Training Objectives vs. M&S [Modeling and Simulation] Capability—which kept the team from focusing on the overall goals. . . . JSIMS died because they did not do a good job defining requirements that meet the stated goals and then mapping them to design.

Special Operations Forces Air Ground Interface Simulator

Air Traffic Control (ATC) and Terminal Attack Control (TAC) remain important training gaps. Extensive training requirements and high turnover rates among trained personnel have resulted in chronic

---

authors of this report "own" limited licenses for copies of Microsoft Office™, but Microsoft retains the right to sales, the proprietary source code, etc.

shortages. To ease the burden and cost of live training (which requires the participation of manned aircraft in addition to ground component trainees), both the U.S. Air Force and U.S. Special Operations Command (SOCOM) sought up-to-date simulations. There was some disagreement over which simulators to pursue, but SOCOM succeeded in getting a joint Operation Requirements Document approved by the Joint Requirements Oversight Council for SAGIS in 2003 (Ashby, 2004, p. 7).

When fully fielded, SAGIS will included a "flexible" database, programming language, and system configuration and will be HLA-compliant. SAGIS will allow the 75th Ranger Regiment, U.S. Marines, and U.S. Navy SEALS to train for call for fire missile, artillery, and close air support (CAS) missions within a joint battlespace (U.S. Special Operations Command, n.d., p. 1).

SAGIS is still being developed and acquired. With congressional approval, SOCOM projects 28 SAGIS training systems nationwide within the next six years (U.S. Special Operations Command, n.d.). SOCOM believes that SAGIS will become the preeminent platform for CAS, TAC, and conventional ATC training within the next six years. (See Figure 3.1.)

SAGIS Business Model

SOCOM requested $9 million for prototype development, but Congress approved only $4.2 million for FY 2004. SOCOM submitted a FY 2005 request for $5.6 million to complete the prototype that could be packaged in transportable containers and that would support mission training, rehearsal, and networking with other simulators (U.S. Special Operations Command, n.d.).

SOCOM's timetable projects three additional systems in FY 2006 and FY 2007 at $2.045 million each, with installation and network setup costs at $600,000 and contractor logistic and instruction support expenses of $810,000 per year (SAGIS, 2003, p. 2).

Although SAGIS is a new system (prototypes were just being deployed as of this writing), the business model appears to be very

Figure 3.1
Sketch of SAGIS Training Interface Concept

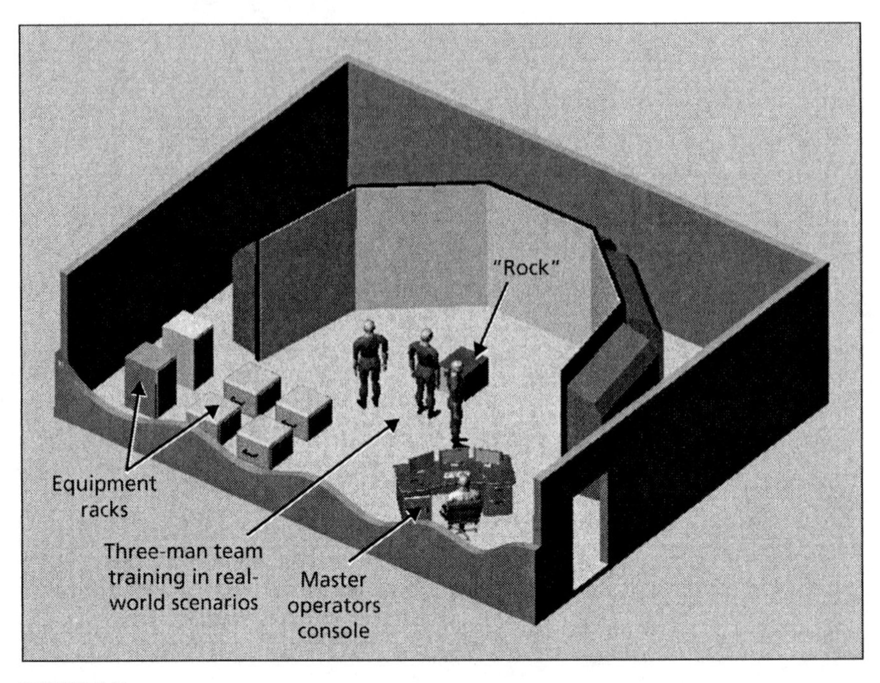

RAND *MG442-3.1*

close to the old way of doing business. The DoD will own the simulations and hold full government rights. Ongoing service and maintenance will come from contractors, likely the same contractors that built the simulators (U.S. Special Operations Command, n.d., p. 2). RAND found no details about procurement contracts for the simulators or the support but expect that they are traditional cost-plus contracts.

The only substantial differences between the SAGIS business model and the old way of doing business are:

1. Many of the simulation tools that SAGIS is composed of are already available as commercial products, which will reduce costs to create the simulators; presumably, SAGIS acquisition contracts specified the use of these COTS products.

2. Training will be delivered primarily by uniformed personnel, so the degree of contracted training support required is likely to be relatively low.

Lessons Learned from SAGIS

The SAGIS acquisition appears to follow the old way of doing business, with two notable exceptions. First, many of the simulators' components can and will be bought as COTS products, resulting in cost savings for the DoD and preventing contractors' make or buy decisions with perverse incentives to "make." Second, and far more interesting, is that SAGIS training will be delivered by uniformed personnel. Arguably, where DoD entities will provide the training support themselves, it makes more sense for the DoD to own the simulators. Other than the fact that uniformed personnel will provide the bulk of the training support with SAGIS, RAND could find no reason SAGIS-like tools could not be leased or purchased on FFP contracts per availability hour or associated with training support services as is done in some of the other business cases considered here.

Virtual Convoy Combat Trainer

Unanticipated resistance by irregular forces and insurgents in Iraq created an urgent need to increase the combat readiness of forces whose roles traditionally were behind combat lines in rear areas, particularly convoy operations and security personnel.

Responses to this urgent need included both traditional training programs and exercises (such as the convoy training program established at Fort Sill, Oklahoma) using real convoy vehicles and live fire (Gourley, 2004), as well as simulation training through such systems as the VCCT.

In March 2004, the U.S. Army Program Executive Office for Simulation, Training and Instrumentation (PEO STRI) submitted a full and open Request for Information (RFI) for a VCCT trainer. The RFI asked firms in the private sector to determine what was commercially available for adaptation to military use by requesting

> information regarding the availability of virtual training systems to be utilized to train drivers and gunners of military vehicles to identify a potential ambush, how to identify improvised explosive devices (IEDs), how to avoid an ambush, how to return fire, maneuver and to react appropriately in the Contemporary Operating Environment (COE) (Gourley, 2004).

The RFI triggered 11 proposals from companies, each accompanied with a technology demonstration conducted in Orlando, Florida, on April 5–9, 2004. A combined team representing Forces Command, Training and Doctrine Command, and the materiel developer, PEO STRI, observed and evaluated the bids.

PEO STRI awarded contracts in June 2004 to two companies, Lockheed Martin Simulation Training and Support of Orlando, Florida ($9.6 million) and Raydon Corp. of Daytona Beach, Florida ($5.6 million). The contractor simulators, the VCCT-L (the Lockheed version) and VCCT-R (the Raydon version), are both mounted on deployable trainers but use different simulation approaches. (See Figure 3.2.)

The convoy trainers provide skill training at the individual- and team-level for the soldier, and include driving, shooting, communications, and decisionmaking in a combat environment. The VCCT "requires soldiers to coordinate actions on a single vehicle, between multiple vehicles and with higher headquarters. This system incorporates precision weapons effects along with driving skills for a variety of vehicles" (Lockheed Martin, 2005).

The Product Manager for Ground Combat Tactical Trainers, Lt. Col. Joseph Giunta, noted that the current VCCTs were intended to provide only a limited capability, since the urgent operational need statement did not represent a fully defined requirement.

"The Army is going to spend the next six to 12 months using this [VCCT] service to help better define the true requirement," Giunta said.

> And at the end of that process, which is being led by TRADOC [U.S. Army Training and Doctrine Command] and FORSCOM [U.S. Army Forces Command], they will either

Figure 3.2
The VCCT-L in Action

SOURCE: Lockheed Martin.
NOTE: Photo depicts the prototype version of the Virtual Combat Convoy Trainer.
RAND *MG442-3.2*

> determine that a requirement is valid and produce a CDD [capabilities development document], or they will say that there's another way of doing this, in modifying an existing device, or they'll just say it isn't a requirement anymore. That's for the Army to determine. (Gourley, 2004.)

With that in mind, the Army started training-effectiveness analysis of both simulator types in April 2005. The Army indicated that new capabilities and functionalities were needed, which were not possible to incorporate initially in the time available.[4]

---

[4] Walker (2004a) notes that these new capabilities and functionalities could include simulation of effects such as night vision, dust, and rear-view vision. The Army also might include other simulations such as close air support aircraft and unmanned aerial vehicles supporting the convoy, to create a total training environment.

VCCT Business Model

Although the VCCTs were solicited from private industry through an initial RFI, the Army did not own the VCCT trainers when they were first delivered; they were leased by the Army, with the option to buy. The vendors that made the trainers retained ownership of both the intellectual property essential to the VCCTs and the actual VCCT simulators on which soldiers train. The Army buys training from these contractors through a fixed-price-per-person-hour of training contract. The contract may specify a minimum amount of training to be purchased or other means of ensuring that the vendor recovers its costs, but the core contract vehicle is FFP.

When the Army awarded the 2004 VCCT contracts to Lockheed Martin and Raydon Corp., PEO STRI rented two mobile suites for the contractors and instructed both companies to provide all training services under the VCCT umbrella. Both contractors leased the equipment to the Army for one year, offering a buyout option. Lockheed offered the eight simulators to the Army for $2 million once the lease expired in June 2005. Six months later, the Army awarded the company a 12-month service contract valued at $4.2 million for a training suite of four VCCTs.

Lessons Learned from VCCT

The VCCT has been a very effective stop-gap training measure. The Army got the "90 percent solution" fielded in a responsive, cost-effective fashion. By allowing industry to bid in response to a loosely specified RFI rather than in response to a requirements-laden RFP, industry was able to be innovative and efficient. By buying training through FFP contracts and leasing the simulation equipment, the Army gave contractors the incentive to continued efficiency and had the option to not buy the simulations.

The business model for VCCT appears to be very similar to Alt#4. The core elements match: FFP training support contracts, the DoD does not own the simulations. Contracts were also short-term, and more than one contract (two versions of VCCT) was awarded. There are some critical differences, however: The DoD did eventually come to own some of the simulations (exercising the option to buy on their leases), and for both VCCT versions, the same contractor provided both the training and the tools. Arguably, the integrated role of contractor as TSP and toolmaker allowed these contractors to respond to the emergent need and be able to fine-tune and perfect their tools even as they were beginning to deliver training.[5]

Also interesting is that two VCCT contracts were awarded and two different VCCT systems fielded. Each is recognized as an effective "90 percent solution" but to slightly different tasks. Going forward, the Army continues to use both VCCT-L and VCCT-R (Blackmon, 2005). From a business model perspective, choosing to award contracts for two systems clearly promotes real competition, both ex and post ante, and cuts in half the DoD's risk of not having any training system. Especially in situations where 80 or 90 percent solutions are sought, multiple awards may prove to be an attractive choice.

Distributed Mission Operations

The U.S. Air Force's Distributed Mission Operations Center at Kirtland Air Force Base, New Mexico, has oversight over the M&S facility for training its warfighters.

The Air Force initiative for simulation-based training at the M&S facility and several other Air Force training sites is called distributed mission operations). The umbrella program emphasizes operational concepts and mission rehearsal by linking aerial warfighters in synthetic combat scenarios. The skills simulators provide real-time training to pilots in mission training centers (MTCs). The MTCs physically house the simulators and other "live feed" equipment that provide stand-alone or distributed training (Brower, 2003). These MTCs exist at Air Force facilities all over the United States and at a few locations outside the continental United States.

---

[5] Blackmon (2005) quotes a Raydon technician indicating that VCCT-R continues to receive updates and improvements: The simulators are updated "as fast as we can update them." The same article reports that soldiers who use the simulators offer feedback and suggestions for improvement, which Raydon uses as input to a continuous improvement cycle.

The simulation models train individual pilots in air-to-ground and close air support operations in either stand-alone or distributed training events (Brower, 2003). The pilots build and test missions while honing their skills with their respective aircraft (Erwin, 2003).

The DMO infrastructure boasts over 175 computers, 20 Local Area Networks, and 15 Wide Area Networks. This computing capability supports an immersive combat synthetic battlespace that models 19 weapons and C4ISR systems.

The DMO concept is expanding from the Air Force to the other U.S. military services. The program is envisioned as a scalable, interservice program allowing for a new level of jointness in air operations. Program managers hope that DMO will be able to train pilots and aircrew in a completely integrated environment by 2007 (Brower, 2003). The goal is to be able to support every conceivable joint training possibility, from the individual all the way up to a campaign-level mission scenario.

The DMO network includes "new" simulators for four aircraft (F-15C, F-15E, F-16, and Airborne Warning and Control System [AWACS]) acquired under the DMO business model. Efforts to incorporate existing DoD-owned aircraft simulators into the DMO network are also under way.

DMO Business Model

The Air Force is currently applying a commercial-fee-for-service contracting approach to new DMO acquisitions (Boeing, 2005). DMO buys hours of simulator availability. It does not buy (or lease) simulators, and training is bought under separate contracts. Contractors provide (and own) the simulators and the upgrades and maintenance for those simulators and are paid a fixed fee per hour for a predetermined number of hours of simulator availability.

The initial acquisitions and modifications to upgrade the simulator software/hardware to DMO requirements have been competitively bid firm-fixed-price contracts. For example, in 1999, Plexsys Interface Products, Inc., an independent small business, was awarded a $75.6 million indefinite delivery/indefinite quantity firm-fixed-price contract to provide simulators for the AWACS MTC program and in 2000 was awarded a $12.6 million modification to provide an upgrade (U.S. Department of Defense, 1999, 2000).

Longer contract terms allow the contractors a better opportunity to recover their investments (Brower, 2003). The contracts have conditional terms, each with a number of "base" years and the potential for performance-based extensions to a predetermined maximum length. One key performance indicator (KPI) contributing to this determination is pilot/aircrew satisfaction with the simulations.

Training is procured separately at the MTC level. Some MTCs run training with uniformed personnel and some contract out for the training. The contracts are let at the Major Command level. Where outsourced, training and simulators are provided by different contractors.

Lessons Learned from DMO

DMO seems to be working very well. User satisfaction is high, and the DoD is paying fixed prices for simulator availability, so there are no cost overruns. It is interesting to note that if simulators are available fewer hours than the contractually agreed amount, contractors are paid less.[6]

Although the first contract under DMO did not pay the commercial training simulation service provider until training was delivered (nearly 24 months after the contract was signed and after Boeing had spent over $1 million), subsequent contracts have paid fixed amounts for demonstrated incremental progress toward training delivery to increase participation by smaller providers less able to shoulder the start-up costs. It worked, as evidenced by the contract won by Plexsys Interface Products, Inc.

One of our interview respondents shared his experiences with the challenges of writing firm-fixed-price contracts. He made three observations. First, it is impossible to plan ahead for everything.

---

[6] The DMO's first contract with Boeing for the F-15 platform provided a number of lessons learned. Initially, Boeing was not able to get the training up and running and did not receive any payment. Once it was operating—only at about 70 percent of target capability—the company was paid that percentage until training was fully functional.

When writing solicitations, DMO personnel included the need to simulate all current features and scheduled upgrades throughout the planned lifetime of the aircraft. However, after 9/11 there was an unscheduled addition of very-high-frequency radios to one of the fighter aircraft. A small additional contract was written to make the change to the simulators.

Second, it is important to solicit bids that make clear what needs to be done without telling the contractors how to do it. To get innovative, best value solutions, do not specify everything as a requirement but instead specify a few core requirements and then a "target training task list" or another expression of desire and needs. DMO works very closely with its contracting office at Aeronautics Systems Command to fine-tune the wording in their RFPs and contracts.

Third, even if everything possible has been included in the contract, there is always the risk that an assumption on the DoD side will be assumed differently by the contractor. For one of its contracts, DMO assumed that the contractor would need to access classified databases to satisfy one of the specified requirements. The contractor assumed otherwise and found a way to satisfy the specified requirement without access to classified databases. DMO had to make a small modification to the contract and pay a surcharge.

The DMO business model involves buying training and tools separately, on FFP contracts, very much like the second step ("an improvement") shown in Figure 1.2. As such, it is a great example of some of the elements of Alt#4 working as designed.

Alt#4 relies on the separation between tool vendors and training providers, as is the case in the DMO model, where it appears to work well. Alt#4 relies on FFP contracts and competition to encourage efficiency and innovation. DMO uses FFP contracts and has received two or more competitive bids for each of the four major simulator service contracts it has awarded. Only two of those contracts went to the same provider (both F-15 contracts went to Boeing, the prime contractor on the full-up F-15 system). The business model elements DMO shares with Alt#4 seem to be working well in the DMO context.

DMO's business model differs from Alt#4 in two important ways: First, DMO focuses mainly on buying simulator hours (the tools side of the equation) instead of training support. However, buying tools as a service for a firm-fixed-price is similar to buying training as a service at a firm-fixed price (as Alt#4 proposes). DMO's approach still allows the DoD to avoid owning these new tools. Second, and a much larger point of contrast, is the long-term contracts used by DMO. If all the performance extensions are realized, DMO contracts can last as long as 15 years. Alt#4 recommends short contracts to encourage competition and keep small vendors' and providers' hopes alive for winning a contract in a subsequent round.

Finally, note that all of the DMO simulators are "heavyweight" simulators—all high fidelity, fully enclosed simulators with a high degree of military specificity. However, DMO has shown that there was competition in the commercial tool market to provide these heavyweight military aircraft simulators on FFP per hours of availability contracts. One imagines that the length of contract offered has something to do with contractors' willingness to bid, and this positive experience bodes well for efforts to expand new and efficient business models into challenging areas, such as heavyweight and exclusively military application simulations.

United Kingdom Ministry of Defence Naval Recruiting and Training Agency

The NRTA provides two cases of simulation training acquisition. NRTA itself is also an example of a wholly different way to organize government entities to provide what are traditionally considered government services (such as military training) in a partially privatized fashion. The United Kingdom's Ministry of Defence established "Defence Agencies" in the 1990s as part of the British government's Next Steps program to improve the effectiveness and efficiency of service delivery to customers.[7] These Defence Agencies are partially privat-

---

[7] The key features of Defence Agencies are:

- the establishment of organizations with clear statements of their roles, objectives, and responsibilities—set out in a Framework Document;

ized to a greater or lesser extent, depending on the specific agency, its purpose and experience, and the results of periodic reviews. Some have been moving further and further "outside" the government. The majority of civilian and military staff at the Ministry of Defence now work within Defence Agencies spread across all functional areas. NRTA remains part of the Royal Navy (RN) with the majority of appointments filled by serving officers and enlisted personnel. However, the roles of many of the staff align more closely with those of a commercial organization and accounts of activity are submitted each financial year in accordance with normal business practice. NRTA is responsible for all recruitment and training activities within the Royal Navy and has an annual budget of $150 million.

Flagship was formed as a joint venture company in 1996 as part of a partnering arrangement with NRTA. The relationship between NRTA and Flagship encompasses three levels of public-private partnership (PPP). Flagship is a strategic partner with NRTA and sells government services (in the form of excess training capacity) in private sector markets. Individual training/simulation programs are competed at the private finance initiative (PFI) level. Flagship competes in and wins many of these PFI bids.

In describing the goal of individual PFI agreements, the Web site of Her Majesty's Treasury (n.d.) notes that

> The Private Finance Initiative (PFI) is a small but important part of the Government's strategy for delivering high quality public services.
>
> In assessing where PFI is appropriate, the Government's approach is based on its commitment to efficiency, equity and ac-

---

- clear and simple lines of accountability, responsibility, and authority;
- the appointment of a chief executive with managerial freedom to achieve best value and personal accountability for the delivery of results;
- individual tailoring of organizations promoting a customer-focused approach to providing services; and
- pressure for improvement through the development of clear and rigorous targets, arrangements for reporting on performance, and appropriate rewards and sanctions.

> countability and on the Prime Minister's principles of public sector reform. PFI is only used where it can meet these requirements and deliver clear value for money without sacrificing the terms and conditions of staff.
>
> Where these conditions are met, PFI delivers a number of important benefits. By requiring the private sector to put its own capital at risk and to deliver clear levels of service to the public over the long-term, PFI helps to deliver high quality public services and ensure that public assets are delivered on time and to budget.

Flagship provides a range of services to the RN across the majority of its training establishments in the areas of facilities management, construction of new facilities, investment in training infrastructure, and the identification and marketing of spare capacity to commercial customers and overseas navies. In some notable instances, Flagship also provides training to other customers using its facilities (e.g., use of the FFTU).

In addition to the FFTUs and MCTS case studies (examined separately), the PPP has led to Flagship providing all facilities management roles at the RN's new entry establishments and major career training schools. Flagship markets RN spare capacity in these establishments and has sold excess capacity for a profit to foreign navies, merchant marine companies, and commercial organizations. For example, Flagship will train Network Rail's engineering apprentices for the next five years at NRTA training establishments for $54 million.

Of prime importance to the success of the PPP is the establishment of joint goals and a joint aspiration to deliver against the "High Level Agreement" that is the cornerstone of the PPP. To achieve this, Flagship and NRTA operate together at all levels across the whole of NRTA's portfolio of obligations. Confidence in the partnership is such that for some projects, RN uniformed personnel are provided to Flagship as part of the individual PFI arrangement[8] and work for the

---

[8] For example, Flagship uses uniformed personnel to deliver some warfare training and pays NRTA the full salary/costs associated with those personnel. NRTA then pays for the service

company to meet both NRTA and other customer requirements. Agreed-on objectives flow down through both organizations with alignment achieved from PFI targets and key performance indicators related to service provision.[9]

The NRTA–Flagship PPP has run for nine years and will be reviewed in 2006–2007 before expiring in 2011. Outcome of the review could range from an extension of the PPP to its disestablishment and termination of the partnership. By all accounts the partnership has been very successful, and representatives from both NRTA and Flagship tout a host of cost savings for the Ministry of Defence over the duration of the PPP.

Fire-Fighting Training Unit

Within the framework of the NRTA–Flagship PPP, a contract was signed in 1999 for Flagship to provide NRTA with unique state-of-the-art FFTUs. This was the first PFI for the PPP and it involved the transfer of all Royal Navy fire-fighting training from facilities owned, maintained (by a managed subcontractor), and operated by the military to those provided by Flagship. The facilities were delivered at three sites in the United Kingdom[10] in 2001. Under the original arrangement, NRTA would use RN uniformed personnel, passed to Flagship in a complex arrangement, to operate the FFTUs. A subsequent business case proposal from Flagship identified financial and other advantages for the partnership if the RN personnel could be transferred permanently to Flagship (as civilians); sufficient suitably qualified RN personnel volunteered to join Flagship to make this proposal work.

The new FFTUs were built alongside existing RN facilities within the naval bases at Portsmouth and Plymouth. Flagship as-

---

provision that Flagship provides at a fixed price, including the support of the uniformed personnel.

[9] Approximately 1,500 KPIs measure service provision performance. Income-generation performance is measured against simple financial targets. Central KPIs include a host of trainee satisfaction measures.

[10] The purpose-built FFTUs are within the naval bases at Portsmouth and Plymouth. Personnel at two small bases in Scotland use local facilities modified by Flagship.

sumed the role of prime contractor by subcontracting to existing manufacturers and system integration engineers the responsibility for providing the specialist equipment and new systems that were needed to deliver state-of-the-art training. Those in Scotland were provided from modified facilities previously used only by local authority fire brigades. These subcontractors have provided their products or services to other customers in other business arrangements, sometimes using their involvement with this project as a marketing tool.

In line with the aspirations of the broader PPP, Flagship has successfully sold the spare training capacity of the FFTUs to a range of customers; the FFTUs are running currently at 105 percent of promised capacity.[11] The profit is split 60:40 between NRTA and Flagship in accordance with the PPP agreement. The non-NRTA customers include foreign navies, merchant marine sailors and companies, and other commercial customers such as a supermarket chain that uses the facilities for management, leadership, and team-building training.

The FFTUs create realistic fire-fighting scenarios, using environmentally safe propane fueled fires and artificial training smoke to achieve an ideal simulated shipborne environment for what is described as a vital element of Royal Navy training. All RN personnel must pass a basic fire-fighting course before being allowed to join a ship for any period longer than 48 hours. Most trainees are RN personnel and are trained to cope effectively with the hazards of fire and smoke at sea, to enable them to prevent loss of life, avoid injury, and minimize damage, thereby enabling the ship to retain its operational capability (Figure 3.3).

Practical instruction is provided in eight highly sophisticated FFTUs. These three-floored propane gas-fueled simulators give the instructors, via a control room, complete control of the fire and environmental conditions faced by students. The compartments within

---

[11] Flagship agreed to make and maintain a certain level of course/simulator availability for NRTA; the simulators have been available more than the planned/promised amount and all that excess capacity has been sold to other customers.

Figure 3.3
Trainees at a Flagship-Owned and -Operated Fire-Fighting Training Unit

SOURCE: Royal Navy and Flagship.
RAND *MG442-3.3*

the FFTUs replicate areas found in ships, such as an engine room, a machinery control room, a messdeck, a galley, and passageways.

Unexpected benefits of the changes to training delivery include

- more efficient working practices and better tools for assessing training efficiency;
- workforce stability and better flexibility to manage manpower;
- evolution of best practice for transfer to other training areas; and
- independent quality control of training standards.

The FFTU PFI is considered a milestone success for the NRTA–Flagship PPP. NRTA believes that it has got better training facilities than it could otherwise have afforded and that RN personnel receive the best training available. Flagship delivered the simulators and other training facilities on time and charges NRTA an agreed fixed-price per training day delivered.

FFTU Business Model

The FFTUs and all associated training are provided by Flagship to NRTA under the terms of a PFI. The normal use of PFIs within the United Kingdom is to allow private industry to provide assets and facilities management to the public sector. In part, this is the case here, with Flagship undertaking the building and maintenance of the simulators and associated training facilities. Flagship's involvement in the operation of the simulators and its ability to market spare capacity to the mutual benefit of Flagship and NRTA make this an unusual and groundbreaking example.

Flagship competed with other prime contractors to win the PFI contract. It sought out suitable subcontractors during and after the competitive process to meet the delivery targets and remains responsible for providing consumables for the operation of the simulators and other associated training facilities. Included in the contract is a requirement that Flagship keep the training and facilities in the fleet up to date with advances in fire-fighting practices. Flagship does this by using the same source contractors as the RN. To date, no major overhaul of the FFTU simulators has been needed.

Flagship is paid a fixed price for delivering training to RN personnel measured in training days. The PFI contract is structured to allow for increases in use above the minimum level to meet operational requirements and these additional days are not charged at an inflated rate. Use below the contract minimum does not result in savings to NRTA, unless the increased spare capacity can be sold by Flagship and the income-generation formula returns money to NRTA. Flagship has full financial responsibility for all aspects of facilities management, service provision, marketing, training delivery, and so on. Some of these costs can be mitigated by income generation according to strict definitions in the PFI contract.[12]

Note that the higher-level PPP between Flagship and NRTA is a very long-term agreement (provided both parties remain satisfied), and the individual PFIs within the context of that PPP are also reasonably long-term contracts.

Lessons Learned from FFTU

Flagship continues to review the opportunities for applying the personnel changes, adopted successfully for operation and training delivery of the FFTU PFI, to other NRTA training areas. This is in accordance with its obligation under the PPP to identify opportunities for improving training and training support efficiencies. At the same time, it is Ministry of Defence policy that service personnel be released from noncore work, such as training, whenever sensible and possible.

Of greater significance, the assessed success of this PFI gave NRTA and Flagship confidence to consider more complex and challenging projects. In particular, the concept for delivery of the Maritime Composite Training System, discussed next, owes much to the experiences gained from delivery of the FFTUs.

The success of the FFTUs serves as an example of success in the United Kingdom's use of PPPs and PFIs to deliver services traditionally delivered by the government, even in the military domain. A similar level of DoD-private partnership could not be realized in the United States without changes in law (recall the close partnership and the flexible assignment of uniformed personnel to Flagship, for example), but the United Kingdom is collecting a body of best practices for this type of endeavor.

Held et al. (2002, p. 36) identify the most common barriers to PPP-like collaborative efforts in the United States:

[12] Because certain details of the PFI contract are commercially sensitive, the description here is deliberately vague.

> The most prominent barriers to greater collaboration are (1) intellectual property concerns, which combines with the fact that most companies do research for their own purposes, not as a service for hire; and (2) excessively bureaucratic requirements and the related distrust of government involvement and oversight in company affairs. When commercially oriented companies weigh these burdens against the relatively small size of the Army market, other limitations on profits, and the perceived fickleness of the government as a customer, the benefits of collaboration generally fail to overcome them.

Maritime Composite Training System

The NRTA, acting on behalf of the Royal Navy, has been attempting to rationalize its training portfolio to reduce costs and improve efficiency. Part of this process has included the relocation of the RN's Warfare Training Center from one site to another. To complete this move and allow the closure of the former site, NRTA needed to move complex, linked Combat Information Center and other combat system simulators while preparing for the introduction and integration of a new class of warship that would include new combat systems.[13]

NRTA and Flagship agreed on a PFI contract in which Flagship assumed responsibility for facilities management and service provision for the existing combat training systems.

The Maritime Composite Training System is an incremental program that builds on the initial PFI with the introduction of new systems and better simulation capabilities. It will allow an integrated approach to all future warfare operator training; all new simulators are required to be compatible with and join the growing network federation. The program has a number of phases, the first of which delivers an enhanced shore training capability and a central warfare hub to facilitate distributed training through a common synthetic environment.

---

[13] The Type 45 destroyer will introduce new radars, missiles systems, and combat management systems to the RN. For more on RN ship procurements, see Schank et al. (2005).

The RN undertakes warfare training at individual, team, and command team or ship levels. Individual skills are taught on career training courses, when, for example, sonar operators learn how to operate each position of a multi-operator piece of equipment. In team training, the focus is to enhance the ability of operators and directors to work together to improve overall effectiveness. Continuing with the underwater warfare example, team training might involve sonar operators manning all positions of the sonar system while senior warfare enlisted specialists are being trained in the direction of the junior operators. Command-team-level training brings together all warfare disciplines and it exercises the ability of the captain, warfare officers, and warfare enlisted specialists to work together using a full CIC simulator.

The courses constitute theoretical and practical instruction and last from two days for refresher training on a specific system to almost a year for warfare officers. The longer courses are modular to smooth out the use of the available simulators and ensure that when team and command-team-level training is needed, trainees can fill all positions.

The scope of MTCS includes the use of the completed Flagship PFI systems that were relocated as well as the provision of buildings to house training and equipment at the two surface fleet bases of Portsmouth and Plymouth. The existing individual and team-level trainers will remain at the new site in Portsmouth. The follow-on phases of MCTS will allow ships in port and at sea to join the synthetic environment and use live systems (with simulated fires) to conduct integrated, federated training.

MCTS Business Model

The relocation project was undertaken by Flagship as a PFI similar to that of the FFTUs, except the majority of the equipment was already owned by the RN and it was retained for use at the new site. MCTS was competed separately and two consortiums came together to bid; one included Flagship in a major TSP role. At the time of writing, the bids are still undergoing evaluation and an announcement is expected shortly. Whichever consortium wins, Flagship will have a place in that business in its role as the PPP partner to NRTA. Flagship also believes that it will be able to pursue further training delivery efficiencies with the transfer of volunteer uniformed RN instructors to Flagship—this is the Training Delivery Business Case proposal.

The main features of the MCTS Phase 1 contract are

- the use of firm prices for provision of initial capability and in-service support, with value for money reviews thereafter at five-year intervals;
- payment in the demonstration and manufacturing stages linked to milestones and earned value management; in-service payments will be made by regular installments in arrears based on annual firm prices and satisfactory availability; and an incentive plan.

The key measure after successful commissioning of the system will be delivery of enough training days to meet the declared requirements. Additionally, the prime contractor will retain full responsibility for the maintenance of the system and will address obsolescence issues throughout the duration of the contract.

Lessons Learned from MCTS

When the MCTS is complete, it will be a demonstration of the U.K. PPP/PFI business model's ability to manage the risks and technological challenges associated with a relatively large federation of networked simulators covering a wide range of training needs. It will also result in a complex series of arrangements between different commercial organizations, each with different contractual responsibilities to NRTA. This MCTS business model is not Alt#4, but is a different innovative approach that may warrant future consideration. See Table 3.1, below, for a comparison and of Alt#4 with case business models.

Central Intelligence Agency's In-Q-Tel

With the advent of the Internet, profit-driven private sector companies accelerated innovation in the technology marketplace, often by investing venture capital in small firms with promising ideas. With the hope of developing and acquiring cutting edge products, the CIA developed its own venture capital corporation in 1999. First called "In-Q-It," it was renamed "In-Q-Tel" in 2000 (Yannuzzi, 2000). Although In-Q-Tel is *not* a business model for the acquisition of simulators or training, it is an innovative approach to stimulating and focusing development in a market comparable to the simulation tools market.

In-Q-Tel was established as an independent nonprofit corporation.[14] It functions like any other corporation, with a board of trustees, professional staff, and business and technology consultants. Its mission is to "foster the development of new and emerging information technologies and pursue research and development (R&D) that produce solutions to some of the most difficult [information technology] problems facing the CIA" (Yannuzzi, 2000). CIA approval is not necessary for any of its business ventures, but In-Q-Tel is held responsible for the outcomes of those deals. In essence, In-Q-Tel is a strategic partner of the government—independent of restrictions and charged with finding real solutions to national security problems.

Yannuzzi (2000) is quick to stress that In-Q-Tel is not a product company but rather a *solutions* company. The corporation does not generate products for CIA use—that is the role of other vendors under separate contractual arrangements.

Success at In-Q-Tel is measured by the return on technology. This success standard depends on whether In-Q-Tel

- "Deliver[s] value to the Agency through successful deployment of high impact, innovative technologies;
- Build[s] strong portfolio companies that will continue to deliver, support and innovate technologies for In-Q-Tel's [Intelligence Community (IC)] clients;
- Creat[es] financial returns to fund further investments into new technologies to benefit the Agency, IC and federal government" (In-Q-Tel, 2005a).

---

[14] It is a 501(c)3 nonprofit corporation, publicly filing with the Internal Revenue Service and operating in full compliance with IRS regulations.

Intellectual property understandings among all parties is important in any research and development agreement. The CIA retains the "traditional government purpose rights" to any In-Q-Tel agreement innovations but allows the corporation and its business partners to retain title to the innovations and to freely allocate any revenues gleaned from intellectual property results (In-Q-Tel, 2005a).[15] This unusual arrangement allows this "hybrid model" of a government/venture capital company to reinvest any profit while keeping the reins of oversight firmly in the grasp of the CIA.

In the five years since In-Q-Tel's inception, it has reviewed almost 5,000 business plans from all 50 states and 25 countries and has delivered more than 100 technology solutions for the CIA and the Intelligence Community (In-Q-Tel, 2005b, 2005c).

As a technology accelerator, In-Q-Tel produces information technology solutions, offering up the CIA as a testbed for technology companies' innovations. An early project called Presidential Information Dissemination System (PIDS) is an example of an implemented pilot. PIDS, an electronic briefing tool for the transition of the president-elect, provides advanced search capabilities and real-time information (Business Executives for National Security, 2001, p. ix). The corporation has invested approximately $16 million in start-up company stock and almost $118 million on technology transfer programs.[16]

In-Q-Tel Business Model

In-Q-Tel is not in the business of acquiring simulation training support or simulation tools but has been included as an example of how a government entity can arrange to disburse venture-capital-like funds to foster innovation in a legal way (as Alt#4 proposes the DoD will

---

[15] The CIA's charter agreement with In-Q-Tel states that "the Federal Government shall have a nonexclusive, nontransferable, irrevocable, paid-up license to practice or have practiced for or on behalf of the United States the subject invention throughout the world for Government purposes."

[16] Eighty-six percent of this $118 million has been spent on contracts with portfolio companies to buy licenses and develop CIA-tailored technology. The remaining amount has been spent on direct equity investments (O'Hara, 2005).

do). The Federal Acquisition Regulations (FAR) contain the rules that control acquisitions and contracting within the Department of Defense and the CIA. Although the FAR serves to prevent waste, fraud, and abuse and a host of other purposes, it brings with it an often unwelcome collection of restrictions and requirements. Business in the private sector can be intimidated by some of the burdens associated with FAR compliance. The FAR also prevents government agencies from spending money in certain ways. To work around these limitations and to encourage smaller companies to work with In-Q-Tel, the CIA developed an agreement for In-Q-Tel based on "Other Transactions" (OT) authority. The OT authority defines the transaction in the negative (as in *not* regulated by the FAR), making it far more flexible. This allows the CIA to fund In-Q-Tel and allows In-Q-Tel to invest in a way that the CIA, on its own, could not.

The CIA's five-year charter agreement with In-Q-Tel includes a relationship framework, policies, and terms for In-Q-Tel contracts. A one-year funding contract is annually renewed. The freedom that this agreement creates allows In-Q-Tel a wide breadth with its business agreements and transactions. It can take the risks that government agencies are loath to do while remaining free of a costly development cycle. As of the summer of 2005, the company's internal rate of return stands at 26 percent, a noteworthy beginning (O'Hara, 2005).

The CIA is the only customer for In-Q-Tel and the private corporation has many advantages over government R&D organizations (Business Executives for National Security, 2001, p. ix): It

- can make equity investments;
- has fewer bureaucratic constraints;
- is not required to comply with FAR requirements;
- is not restricted by civil service personnel policies;
- engages only in unclassified projects;
- has the cachet of being associated with the CIA; and
- has a flexible deal structure modeled after commercial contractual/investment vehicles.

Lessons Learned from In-Q-Tel

In 2001, a nonpartisan, nonprofit organization called Business Executives for National Security, made up of business leaders, assessed In-Q-Tel and provided some recommendations along with cautious praise. The panel found that the In-Q-Tel model needed to mature and did not recommend expanding In-Q-Tel's customer base beyond the CIA (Business Executives for National Security, 2001, p. v). Most of the investments that In-Q-Tel has made have not been "cashed out" at this time, so the corporation could yet yield negative numbers (O'Hara, 2005).

Expectations of In-Q-Tel's ultimate success or failure differ. The main point for this analysis is that In-Q-Tel has been legally disbursing funds in a venture-capital-like way on behalf of the CIA and thus stands as a possible example of how seed money/venture capital might be disbursed under a broader Alt#4 business model for the DoD.

The Alt#4 Model Has Points in Common with Other Approaches but Also Some Distinctive Characteristics

Although In-Q-Tel is an example of one way a government entity can legally engender venture capital support for innovative technologies, the other six cases represent different business models for the acquisition of simulators and simulation training. Each model differs from the others, and each differs from Alt#4. This section compares and contrasts the key differences across business models.

Table 3.1 characterizes each business model on nine variables of interest. The table entries are based more on the abstract qualities of the business model than on the actual "real" case, which results in some discrepancies; these are intentional and illustrative.[17] The nine identified variables of interest are

---

[17] For example, no training was ever delivered under JSIMS, yet we have discussed how it likely would have been had the JSIMS business model not failed. Similarly, as of this writing, the U.S. Army has exercised the buy option on some leased-with-option-to-buy VCCTs. The business model did not require that the Army own those simulators; it gives an example of a DoD entity buying delivered training only and *not* simulation tools.

Table 3.1
Features of Different Business Models for the Acquisition of Simulation Training

Model	Who Buys Tools	Who Funds Tool Development	Who Builds Tools	Who Owns Intellectual Property [*or Assembled Simulators*]	"Units" Tools Provided in	Who Provides Training	"Units" Training Provided in	Length of Contracts
JSIMS	DoD	DoD	Contractor	DoD (full government rights) [*DoD*]	Cost-plus	Same contractor	Billable contractor hours	Long and locked in
SAGIS	DoD	DoD	Contractor	DoD (full government rights) [*DoD*]	Cost-plus	Uniformed personnel	Classes	Acquisition length for tools, no contracts for training
VCCT	Contractor	Contractor	Contractor	Contractor [*Contractor*]	(a)	Same contractor	FFP person-hours of training	Relatively short
DMO	DoD	Contractor	Contractor	Contractor [*Contractor*]	FFP hours of simulator availability	Uniforms or different contractors	Varies	Long, performance extended tools contracts
FFTU	PFI contractor	OEMs	OEMs, subcontractor	OEMs [*PFI contractor*]	(a)	PFI contractor (with transferred MOD personnel)	FFP training days; excess capacity sold for MOD & PFI profit	Long
MCTS	PFI contractor	OEMs	OEMs, subcontractor	OEMs [*PFI contractor*]	Fixed-price for tool availability	RN uniforms and PFI contractor	FFP training days	Long
Alt#4	TSP contractor	OEMs, perhaps w/ catalog conductor seed money	OEMs	OEMs [*TSP contractor*]	(a)	TSP contractor	FFP per training outcome	Short

[a] Blank cell indicates that the training user, DoD or MOD, is not buying tools but is instead buying training outcome. The tool purchases are up to the contractor providing the training.

- Who purchases tools? Tracks whether the simulation tools are bought directly by the defense entity (DoD or MOD) or by a contractor and then furnished to the defense entity as part of a contract for training directly.
- Who funds tool development? Records the source of funding for any tool development, innovation, or integration necessary before delivery. Development is funded either by the defense entity, by a contractor, or independently or indirectly by the OEM, a generic term to describe a commercial simulation tool vendor.
- Who builds the tools? Reports who builds the tools under the business model.
- Who owns the intellectual property associated with the tools? Identifies the owner of the right to modify or sell future instances of this tool, as opposed to . . .
- Who owns the assembled simulators? Who owns the simulation tools used to provide training in this program?
- What "units" tools are provided in? On what basis are the tools made available to the government? This variable is assessed only where tools are made available to the government without that availability being contingent on a broader contract for training support. How exactly TSPs that provide training (and implicitly tools) acquire those tools is not a core concern of any of these business models.
- Who provides training? Reports whether uniformed defense personnel or contractors provide the training with the simulation tools, and whether the same or different contractors provide goods or services elsewhere in that case.
- What "units" is training provided in? Indicates how training is provided/purchased, either in billable hours (cost-plus), firm-fixed-price per given unit (training days or training outcome), or by uniformed personnel at nontransparent costs.
- What is the length of contracts? Indicates whether contracts for simulation tools or training are short (one to three years), of moderate length (four to five years), or long (six years or longer). Note that observed contracts are only short or long.

Table 3.1 highlights several key differences between the various business models. We wish to highlight several of them to consider the consequences of these differences in light of economic theory and literature (discussed in the next chapter).

The first critical difference is in what the defense entity buys. The DoD (or MOD) variously buys simulation tools, simulation training, tools and training, or, in the cases of DMO, FFTU, and MCTS, *availability* of simulation tools. The core conceptual difference is between buying tools as goods (and then owning them) or buying tools as a service (either directly as a service or as part of a training service package).

Second, there are critical differences in the terms of the contracts under which the defense entity buys whatever it buys. Contracts are either cost-plus contracts or firm-fixed-price contracts, and these contracts are either short or long.

The third critical difference is whether the simulation tools and the training come from the same contractor/provider; in several of the business models, delivery of training is wholly or partially integrated with provision of tools; in some of the models, tools and training come from different vendors.

Fourth, funding for tool development covers the spectrum from public (funded by the DoD) to private (funded by contractors, tool makers, or OEMs themselves). All the case study business models fall near one of the two extremes on this spectrum. Alt#4, however, offers the possibility of hybrid funding, where the main responsibility for funding tool development lies in the private marketplace, but the DoD can contribute to tool development through seed money investments. All four of these critical differences receive attention in the next chapter.

CHAPTER FOUR

The Economic Underpinnings of Alt#4: Relevant Theory and Literature

Chapter Two presented Alt#4 and its logic. Chapter Three presented summaries and described business models of other ways of buying simulators and training. Taken together, the two chapters leave a handful of open questions regarding the relative merits of different ways to do business. This chapter considers these questions in light of economic theory and research.

This chapter reviews economic theory and literature for four areas that are either central to the logic of the Alt#4 business model or represent key difference between Alt#4 and other business models. They are

- ownership of tools;
- cost-plus versus firm-fixed-price contracts;
- tied or untied markets; and
- competition and innovation.

Each is discussed at length in its own section. Implications for Alt#4 follow at the end of the chapter.

Ownership of Tools: Contract Length and Investment Incentives

This section considers the implications of owning tools versus renting them (buying tool availability as a service or as part of training services is "renting" to economists) and the incentives that go along with contract durations. One key feature of Alt#4 is that the DoD will no longer own training tools but will instead rent them indirectly as part of a training contract.[1] This restricts the use of Alt#4 to training areas in which tool vendors are willing to provide tools without selling them to the DoD. Theory suggests that where tools would need to be extensively tailored to DoD needs, vendors will be hesitant to just rent or license them. The risk they face is known as the *hold-up* problem. The more customized a product is to DoD needs, the more valuable it is to the DoD but the less valuable to other users. Whoever needs the product is vulnerable to subsequent demands to renegotiate price or use—in other words, to being "held up." The transaction costs theory associated with Williamson (1979, 1985) argues that if contracts cannot be specified clearly and exactly ex ante, and if ex post problems are likely to arise, then one solution is vertical integration, or ownership of the asset. A firm that cannot function if it is held up will tend to seek to own the specific asset outright.

Another implication of the hold-up problem is that a firm will be less likely to invest in such a specific asset if it perceives insufficient benefits from the investment in the future. The DoD's experience with DMO suggests that some vendors are willing to provide very complex military-specific simulations (cockpit simulators) to the DoD on what is effectively a rental basis. One key to this appears to be the length of the contract.

Transaction costs theory argues that vertical integration or long-term contracts can solve the problem of ex post opportunism regarding specific assets in certain contexts. Joskow (1987) provides empirical evidence from coal markets showing that the more specific the

---

[1] From a private sector accounting perspective, DoD is moving from an up-front cash outlay for ownership (capital investment with noncash depreciation charges over the economic life of the asset for tax and financial reporting) to a cash outlay for each year of rental use. When renting, cash outlays are spread over time. From a DoD appropriations standpoint, the outlay moves from a current year cost and is spread over future years. Most likely, the outlay also moves from a procurement appropriation to an operations and maintenance appropriation. This report assumes that there are no exogenous reasons this change of appropriation type could not occur.

asset, the more likely buyers and sellers are to negotiate longer-term contracts. The most prominent example of this can be seen in a comparson of American and Japanese auto manufacturing.[2] American procurement relies on short-term, lowest-cost bidding from suppliers held at arm's length on fixed-price contracts. The buyer inspects the delivery and can reject the order if it does not meet quality standards. Specialized, complex items are produced in-house and if contracted out, the automaker owns the designs and even the parts used by the supplier (e.g., dies) to manufacture the item.

In the Japanese model, automakers develop long-term relationships with fewer suppliers and place smaller and more frequent orders. In turn, the supplier produces the designs for the parts and owns any assets required to produce the part. The automaker does not inspect the delivery but instead monitors quality, costs, and cooperativeness and ceases to do business with an underperforming supplier. An underperforming supplier soon finds that a bad reputation harms its opportunities with other automakers.[3] The contracts are not long and do not specify details; rather, there is a shared understanding by both parties that they will continue to work together over the long run and problems will be worked out along the way. There is little turnover in these arrangements, so although the contracts are short, the relationships are not. Taylor and Wiggins (1997) show that either system can work but that as inspection costs rise (for example, because of increasing asset complexity), the Japanese system is relatively more efficient. The Japanese model also most closely approximates the system that characterizes defense procurement in the United States.

Current trends in commercial supply chain management are moving toward long-term, partner-type relationships. Two successful

---

[2] This discussion is drawn from Holmstrom and Roberts (1998) and Taylor and Wiggins (1997).

[3] A supplier's need to maintain a good reputation can at least somewhat reduce any incentive to engage in ex post opportunism. This is supported by empirical evidence in the defense and software industries as well as in the automotive example (Crocker and Reynolds, 1993; Banerjee and Duflo, 2000).

case examples, DMO and the U.K. MOD's various efforts, rely on different forms of long-term contract. Although different, both rely on the idea that a long-term, accountable relationship allows contractors a chance to realize the profits from their investment while allowing defense partners to shift some of the risk and responsibility to contractors. In contrast, the Alt#4 business model proposes to recompete contracts often to encourage innovation and efficiency through competition. Transactions costs theory suggests that Alt#4 will be most successful in training contexts that do not require training tools highly tailored to DoD needs. However, the possibility of long-term relationships without long-term contracts (following the Japanese auto industry model) may allow Alt#4 to succeed in technology areas in which theory suggests it should struggle.

In areas that do not require extensive investment in highly DoD-specific tools, if there is enough competition, contract length is unlikely to matter. A TSP that is unsuccessful in bidding for a short-term training contract in one functional area could be successful in another. TSPs need only win enough short-term contracts to be successful in the long term. Where tool providers will be providing lightweight simulations using technical means known to them (primarily sunk costs), a few uses of the tool with rapid short-term spiraling to improve tools could meet their long-term sustainability goals. However, short-term contracts are unlikely to encourage vendors to make tools that require extensive investment to develop and are DoD-specific.

Economic theory suggests two central benefits that the DoD stands to realize from Alt#4s proposal to get the DoD out of the business of owning simulation tools. The first is the reduction of switching costs, and the second is increased access to tools from nontraditional military suppliers (NTMS).

Switching costs are those incurred when switching from one good, tool, or provider of goods or tools to another. Switching costs are discussed further below when contract types are compared. Switching costs are much more transparent in the tool ownership case. Simply put, if the DoD already owns something and has paid for it, the relative cost of switching to a newer, better tool will be quite high. However, if the DoD is renting a tool, the cost of switching to a newer, better tool may still not be zero but will be much lower.[4]

In addition to getting the DoD out of the business of owning simulation tools, Alt#4 explicitly gets the DoD out of the business of contracting directly with tool vendors. One benefit of having TSPs buy or license the simulation tools is that the DoD can end up indirectly gaining access to tools that might not have been possible to own because of the reluctance of NTMSs to do business with the DoD.

Lorell and Graser (2001, pp. 10–11) note:

> In the late 1980s and early 1990s, a large number of studies conducted both inside and outside the government concluded that the maze of special government laws, regulations, reporting requirements, and policies imposed on contractors doing business with the government had created two serious problems. First, compliance with the laws and regulations by firms, combined with the extra cost of mandated government monitoring and oversight activities, had resulted in a significant cost premium added to items procured by the government. Government regulations often require that companies comply with hundreds of costly and time-consuming reporting rules as well as with similar government-unique accounting and socioeconomic requirements. According to studies conducted at this time, government regulation increased costs to the government by 5 to 50 percent.
>
> Second, AR [acquisition reform] advocates claimed that government-mandated procedures and standards often have not been in conformity with routine DoD Regulatory and Oversight Compliance Cost Premium business practices in the commercial world—as a result of which many commercial firms have consciously avoided doing business with the DoD.

[4] Switching cost is the difference between the cost (or rent) of the new tool and the ongoing costs (or rents) of the old tools. Where old tools are owned (sunk costs) with low or no ongoing costs, this difference is maximized (so switching cost is at its highest).

Held et al. (2002, p. 36) report that NTMSs are often reluctant to do business with the DoD for a variety of reasons. Untying training and tools and having the DoD transact only with the TSPs should remove many of these barriers, because the tool vendors have business relationships directly with the TSPs and not at all with the DoD.

Cost-Plus Versus Firm-Fixed-Price Contracts

Traditionally, the DoD buys training and simulations through cost-plus contracts. Alt#4 proposes to use fixed-price contracts instead. Each contract type has both benefits and costs, as discussed by Bajari and Tadelis (2001) and summarized in Table 4.1. They consider cost minimization, flexibility, and quality incentives under each contract type. They do so by formally modeling a firm's decision to make or buy a product.[5] This decision depends on the costs associated with product design: the buyer's impatience and the complexity of the product.

Firm-fixed-price contracts dominate when cost minimization is the primary consideration. A cost-plus contract has similar incentives to making the product internally. Because all costs are covered, there is less incentive to search for cost efficiencies than there is under fixed-price contracting. The fixed-price contract, in contrast, has incentives similar to buying the item from an external supplier. Because the supplier's profit depends on maximizing the difference between

[5] The authors model the make-or-buy decision as a choice between two contract types: fixed-price and cost-plus. Cost-plus contracts are considered analogous to a decision to rely on internal production (i.e., to make a product), and fixed-price contracts are analogous to the buy option. Bajari and Tadelis argue that procurement is generally characterized by adaptation over time, as initial designs fail or regulations or needs change. Thus, they model contracting in two stages. In the first stage, a buyer designs the product (specifies needs). In the second, a supplier provides the product. The more complete the first stage, the less complicated the second stage and the fewer changes will likely be required to the contract. Completeness can be expensive, however, as can impatience on the part of the buyer. The product is then provided by the supplier, who can determine non-contractible ways to reduce costs.

Table 4.1
Fixed-Price and Cost-Plus Incentives

Outcome/Incentive	Fixed-Price	Cost-Plus
Risk allocated to	Contractor	Buyer
Incentives for quality	Less	More
Buyer administration	Less	More
Minimizes	Cost	Schedule
Documentation efforts	More	Less
Flexibility for change	Less	More
Adversarial	More	Less

SOURCE: Bajari and Tadelis (2001).

the amount paid by the buyer and the amount it costs to produce the item, the supplier has an incentive to search for cost efficiencies. Moreover, the buyer's need to invest in close cost supervision and contract monitoring is lower in a fixed-price setting than in a cost-plus setting.

However, buyers almost always have additional considerations. Bajari and Tadelis show that when time to completion is an issue, cost-plus contracts minimize the amount of time it takes to complete a project. Changes to contracts are also much more contentious under fixed-price contracts than they are under cost-plus contracts, as any change to the scope of work will affect costs and thus the supplier's profit margin in a fixed-price context. In a cost-plus setting, changes are more easily accommodated because the contract specifies in advance that all costs will be covered by the buyer. It is true that under either contracting type, renegotiating contracts can cost both time and money; such negotiations are less likely under a cost-plus contract.

For a number of reasons, there might be changes required after a contract is executed. Increases in volume (need to train a larger force than anticipated), respecification of needs (as a newly emerged training need continues to evolve), or failure to specify an important goal or requirement can all lead to changes. Generally, the more innovative a product (i.e., something that has not been done before) or the more complex it is (i.e., the more difficult or costly it is to specify exactly what the buyer wants in advance of signing the contract), the harder it is to correctly specify a fixed-price contract and the more likely it is that the contract will need to be amended or renegotiated. Bajari and Tadelis find that a cost-plus contract is more effective when there is a high probability of design failure or greater uncertainty. Cost-plus contracts also give suppliers greater incentives to invest in quality improvements because all of the costs associated with the improvement are covered.

However, the open-ended nature of cost-plus contracts, in which total costs are not specified in advance, gives suppliers an incentive to underbid to win the contract and then engage in noncompetitive behavior after winning the contract. This procedure is common and has led to recommendations that such contracts be small and rebid frequently, as repeated bidding can create competitive pressure. Williamson (1979) observes that frequent contractual relations can curb opportunistic behavior on the part of the contractor. Whether this can be accomplished in practice will depend on the nature of the good under consideration. Contractors may balk at short contracts for goods for which they will require long relationships to recoup their initial investments (a category that may include some heavyweight simulation tools).

Another strategy to avoid underbidders is to switch to fixed-cost contracts once a product is well established. Crocker and Reynolds (1993) studied Air Force engine procurement contracts. They found that early stages of product development were covered by cost reimbursement contracts and that changes to the scope of work were expected and, indeed, occurred. Later acquisitions were procured through fixed-cost contracts, after the engine technology had been well developed.

There remains the issue of "asset specificity," that is, the extent to which a good is customized to the customer. Williamson (1979, 1985) finds that the decision to make a good internally (analogous to cost-plus contracting in this discussion) is more likely the more customized a product is, when fewer substitutes are available in the market. A cost-plus contract is more forgiving of the give-and-take process required to design and build a product to a buyer's exact specifications when the buyer cannot describe fully what is necessary and what is desirable in advance of writing the contract.

Table 4.2 summarizes when a fixed-cost contract and when a cost-plus contract is more likely to be appropriate to a situation, according to current economic wisdom. Fixed-price contracts should work best when the buyer is able to specify what is needed in advance of writing the contract and when the product desired is a more standard, off-the-shelf solution. To the extent that training situations under Alt#4 fit that description, the use of fixed-price contracts should present no problems. When requirements are hard to specify initially *or* the end asset is customized and likely to be useful only to the DoD, economic theory expects cost-plus contracts to work better.

In situations where FFP contracts are likely to be problematic, one option might be to implement performance-based contracts. In this case, remuneration depends on how well the contractor performs. Theoretically, optimal performance metrics have the following characteristics: quantitative (objective), visible (transparent), understandable, multidimensional, comprehensive, aligned with objectives, targeted, cost-effective, and trust-inducing (Gibbons, 1998). Defining optimal metrics is difficult, however. Consider in this case that the metric must measure the outcome (or outcomes) of interest to the buyer. When fixed-cost contracts are difficult because the buyer cannot specify needs clearly in advance, it is also likely to be the case that the buyer cannot determine exactly what metrics will define adequate performance. Moreover, it is easier to measure inputs than outputs, to

Table 4.2
When Might FFP Contracts Be Appropriate?

	Ability to Fully Specify Demand Ex Ante	
Asset Specificity	**Low**	**High**
Low	Cost-plus	Fixed-price
High	Cost-plus	Cost-plus

measure effort rather than achievement or performance, and this will be even more true when buyers cannot specify those outcomes clearly. This is antithetical to the idea of performance-based contracting. It might make more sense in this case to rely on cost-plus contracts and recognize from the start that product design will be negotiated over time. In short, theory suggests that the Alt#4 business model will be most appropriate and most successful in areas where training user needs can be clearly articulated and the training tool is not extensively customized to DoD requirements.

Untying Tools and Training Markets

Alt#4 is predicated on the view that one problem with the old way of doing business was that traditionally the same firm provided the training tools as well as the required training. In the economics literature, this is known as the problem of tied markets: A firm with monopoly power in one market uses this power to monopolize a second market. The two goods can be complementary, as in this case, with the technology used in training and the training itself, or more generally with follow-on maintenance and support services that accompany the purchase of a good.

Before the contract is signed, the market for training can be competitive and firms will compete to land the contract. Once the contract is signed, however, the buyer may incur high costs were it to switch to another supplier. In essence, the buyer is locked in, and an ex post monopoly develops with monopoly prices being charged. These monopoly rents may or may not be competed away in the ex ante competition, but even where they are, the opportunistic strategies that suppliers engage in to land the contract can distort buyer choices (Farrell and Klemperer, 2004).

In addition to the switching costs noted above related to ownership of tools, switching costs can occur when there is some value to an established long-term relationship between a buyer and seller. This could be because the supplier has invested time and effort learning what the buyer needs and anticipating future demand, and the buyer finds that negotiating with a supplier that already knows how it operates lowers the cost of procurement. As the two parties develop specific information about the other, their negotiations in the future are smoothed. This proved to be the case in the NRTA–Flagship PPP discussed in Chapter Three.

Williamson (1985) elaborates on two kinds of specificity, site and asset. In some industries, having suppliers located nearby or in strategic locations (i.e., site specificity) can save costs on transportation of material. Asset specificity is perhaps the more relevant factor in the case of training, in which a supplier designs or customizes products to the buyer's requirements. Heavyweight flight simulators would fall into this category. The more specific the asset, the more likely that the buyer and seller have collaborated over time in the design process, articulating needs and outlining capabilities. In the process, it becomes less desirable and more difficult for buyers to switch suppliers, and a bilateral monopoly is created.

Because of the potential for the exploitation of monopoly power, tied markets have been the focus of a considerable amount of antitrust litigation in the United States, beginning in 1957 with a Supreme Court decision regarding Eastman Kodak copiers. Independent copier repair firms had sprung up to service Kodak copy machines, purchasing spare parts from Kodak. As the copier market plateaued and the service market grew, Kodak subsequently refused to sell parts to the independent firms, claiming that it made them too competitive with Kodak's own copier service work. The Supreme Court declared that Kodak had engaged in monopoly behavior, in essence by tying markets. Kodak was ordered to sell parts to the independent repair firms and, indeed, the standard solution that evolved starting with this case was to break the market apart. This is what Alt#4 proposes to do by barring firms that provide training from also providing training tools.

However, tied markets may not always be a problem. Tirole (2005) argues that tied markets are not in and of themselves evidence of noncompetitive practices and opportunistic behavior and that there can be legitimate benefits to tying markets as well. Lower distribution costs and lower transaction costs may justify integrating products. Tirole gives the example of Peugeot selling bicycle saddles, brakes, wheels, and other equipment as a single assembled product—a bicycle. Although one could buy a Peugeot frame, a Campignola drive train, Continental tires, etc., and assemble a bicycle, the total cost per value would be higher than just buying a whole bike from Peugeot because of the efficiencies in Peugeot's production and integration process.

Product integration and compatibility might also be less costly when markets are tied. This certainly has been one of Microsoft's arguments for incorporating new applications into updates of its Windows operating system, and it has merit even considering allegations of monopolistic behavior on Microsoft's part. Intellectual property might also be protected more adequately when products are tied if, to achieve compatibility, one firm must reveal the proprietary knowledge underlying a good. Again, Microsoft comes to mind, with its resistance to making its underlying code open knowledge, but this is true generally in the computer software industry where interoperability is more talked about than accomplished.[6]

Under Alt#4, the DoD stands to lose the "economies of scope" (e.g., lower transaction costs and product integration) that come with tied products but stands to gain if competition can drive prices down and increase innovation. There will need to be competition among TSPs for DoD business and between vendors to provide tools for the catalog from which trainers choose. The process works in tandem. The more user-friendly the tools, the more likely that more than one trainer will compete in the market because the barriers to entry will be lower. There need to be multiple tools to choose from. The switching cost problem does not disappear if trainers also find it very costly to switch tools, but then at least the problem is the TSP's, and not the DoD's. If the TSP becomes noncompetitive because of its refusal to pay switching costs, the DoD is free to contract with its competitors. Varied catalog options will in turn be more likely if the

---

[6] Tirole (2005) also lists other potential advantages to tying, including information and liability considerations, legitimate price response, and market segmentation, but these are less relevant to the defense procurement issue under discussion.

DoD is able to attract nontraditional vendors (such as the computer games industry) to develop and offer tools for certification, if the standards and the certification process are not overly burdensome and the venture-capital-like activity effectively seeds product development as needed.

The firewall between trainers and firms that develop training tools is a key feature of Alt#4, but it may not be the optimal arrangement for all types of simulation tools and training. Where it is, it may not be necessary to maintain it over time. All training tools must pass an objective certification process and so it should not matter if trainers select an approved tool that their own company has developed in the same way another trainer might choose that same tool. The certification process should ensure that any trainer is able to use any product listed in the catalog (in other words, that the tool is equally user-friendly to firms that are not associated with its development). Then the firm that developed the tool will not have an undue advantage in the training market and will not be able to exploit its position in one market to dominate the other. It should become apparent over time if the product development and certification process is sufficient to protect against tied markets, and the firewall could be relaxed.

Competition and Innovation

Alt#4 includes seed money investment to encourage innovation and competition in the tool vendor market. Birkler et al. (2000) find that existing DoD acquisition approaches are not optimal for defining and developing innovative or novel system concepts. Traditional DoD acquisitions, they argue, have too long a cycle time and are too risk-averse to be conducive to innovation.[7] Commercial firms, on the other hand, have repeated success stories for rapid innovation. Within

---

[7] Along similar lines, Rogerson (1989) argues that defense firms must be allowed positive profit, i.e., profit that does not get competed away, to fund investments in innovation.

the last decade, venture capital investment has been the champion for efficient funding of innovation.

Using patenting of inventions as an indicator of innovation, Kortum and Lerner (1998) found that a dollar of venture capital was five to 14 times more effective than a corporate R&D dollar in terms of innovation. Assuming that defense R&D dollars are comparable, the venture capital investment is a marked improvement. Held and Chang (2000, p. 1) note that "studies have also verified a positive correlation between venture capital and innovation, noting for example the high rates of patenting activity for firms backed by venture capital and the large R&D investments these firms make in comparison to other companies."

Economic theory suggests that the relationship between innovation and competition is positive, up to a point; when competition is too fierce, however, innovation suffers as competitors are forced into thinner and thinner profit margins, leaving them limited finances for research investments and reduced risk tolerance for the possibility of failed innovation efforts (Held and Chang, 2000). The overall relationship between innovation and competition is shaped like an inverted "U" with competition increasing innovation up to a point beyond which competitive pressure acts as a damper on innovation (see Figure 4.1). In general, competition can be viewed as a positive for innovation, barring this theoretical limit. Given the DoD's traditional problems in generating sufficient competition, the threat to Alt#4 from excessive competition is not considered a serious risk.

Under Alt#4, seed money is used in a venture-capital-like fashion either to directly encourage innovation or to encourage competition, which should also encourage innovation as well as help prevent monopoly power in the tool market. Both venture capital investment and competition have been shown to contribute to innovation.

Figure 4.1
Competition Increases Innovation, Unless Competition Is Extreme

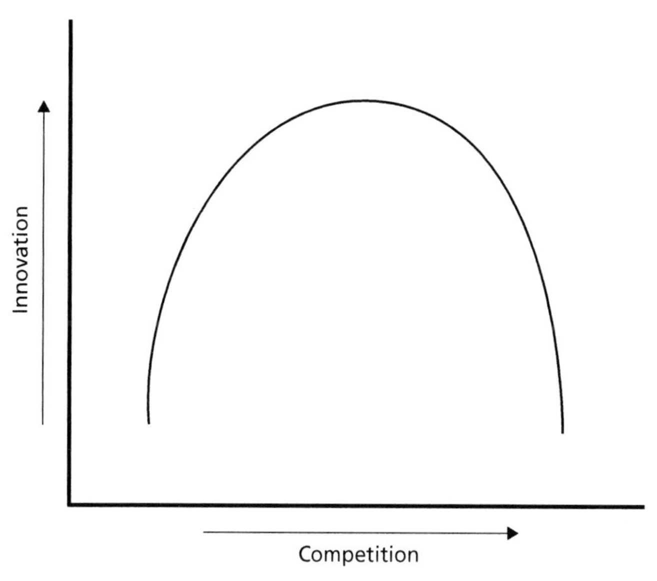

RAND *MG442-4.1*

Summary

Table 4.3 summarizes the key points from economic theory that pertain to Alt#4. Note that every key aspect of the Alt#4 business model has some risk or possible drawback associated with it, but then so do all of the alternatives Alt#4 replaces. Alt#4 represents a different set of choices and tradeoffs than the old way of doing business and requires a different set of risk-mitigation strategies to realize optimal performance. Specific risks and mitigation strategies are discussed in the next chapter.

Table 4.3
Summary of Key Findings from Economic Theory

Topic	Benefits	Limitations and Risks
Owning tools	More highly tailored product; less vulnerable to supplier opportunism (hold-up)	Responsible for maintenance and updates
Renting tools	Supplier/owner responsible for maintenance and updates	Supplier might force renegotiated terms by holding up a key product
Long contracts	Cooperative relationship can develop	Fewer opportunities for competition; participants "locked in"
Short contracts	Competition maintained; losers of previous competitions have frequent opportunities to bid again	Fewer incentives for contractors to innovate in DoD-specific tools; less certainty of "relationship"
Cost-plus contracts	Flexible enough to adjust when needs are difficult to specify at outset; incentive to invest in quality improvements	Fewer incentives to contractor efficiency; high monitoring cost
Fixed-price contracts	Incentives to contractor cost efficiency; reduced monitoring cost	Quality disincentive; changes in contract difficult, possibly adversarial
Untied markets	Increased competition	Intellectual property issues; loss of economies of scope
Integrated markets	Can be cost-efficient; tool vendors may know best how to use their own tools; product compatibility incentives	Firm with monopoly power in one market may use it to dominate a second
Competition	Increases incentives to efficiency; lowers costs; promotes innovation	Extreme competition can hamper innovation

Implications for Alt#4

This chapter has considered the economics and business literature on contracting, tied markets, and ownership in light of the characteristics of Alt#4. The main conclusion we draw from the economics literature is that Alt#4 is based on sound economic principles and has good prospects for delivering efficiencies to DoD's training community. Our second conclusion is cautionary; theory suggests that the Alt#4 model is most likely to realize the cost efficiencies and innovations of the private sector if it is applied to technologies that also commercial applications and training needs that are relatively straightforward to specify. Theory suggests that the more tailored training must be to DoD specifications, the less likely it is that many firms will bid on work developing training tools (because there is likely to be no profitable commercial spillover). The extent to which these specific risks concern us is mitigated by the successful example of DMO, presented in Chapter Three, above. The caution remains: There may be certain areas of training or of simulation tool technology for which Alt#4 may not be the best business model, and any future effort to expand the application of Alt#4 should move cautiously to make sure the model is appropriate to the tasks to which it will be applied. Alt#4 may not be a "one size fits all" solution.

CHAPTER FIVE

Challenges Facing the Prototype

The preceding chapter considered the core logics of the Alt#4 business model and in light of economic theory and experience. Although the logics Alt#4 relies on have sound foundation in economic theory, they are not wholly without risk and are not always clearly better than approaches taken by other business models. The right choice of business model depends on the economic reality of the situation in which it is applied. Having a prototype and opportunity to see how the logics of Alt#4 function in practice remains a good idea. Based on the economic analysis and on RAND's observations of the acquisition and prototype implementation context, this chapter presents identified challenges and risk areas for the successful implementation of an Alt#4 prototype. Note that these challenges and risks are not all of the same magnitude, but any of them, if not dealt with effectively, could impede the success of the prototype. The discussion includes risk-abatement strategies relevant to those challenges where possible.

Standards Setting

Although not presented as a particularly difficult problem in the TC AoA final report, getting the compliance standards for entry into the Alt#4 simulation tool catalog right is critical and challenging.

- Standards are critical to several aspects of the Alt#4 business model.

- Standards ensure that tools, in the aggregate, are moving in the direction of L, V, C integration that serves the DoD's long-term training goals.
- Standards ensure that tools purchased by TSPs from the catalog will work and will work with other tools purchased in the catalog.
- Standards prevent monopoly power in the tool market, by preventing tool makers from entering tools into the catalog that use proprietary protocols. If modular simulation tools are modular only with respect to other tools made by the same vendor (and running on the same proprietary protocol), then the ability of other vendors to compete with their own modular tools is compromised, as is the ability of TSPs to mix and match and plug and play to deliver the best quality for best value training that is the core objective of the Alt#4 model.

Limited time prevented us from fully investigating the challenges associated with standards setting, but we did hear from experts we spoke to of several anecdotes regarding the failure of standards. These accounts made it clear to us that standards-setting is a challenge, as did a recent RAND study by Davis and Anderson (2003). In setting standards, a balance must be struck between exclusivity and permissiveness. Standards must not be so difficult to meet that they exclude good tools, innovative ideas, or nontraditional military vendors. However, standards must not be so simple to satisfy that vendors can certify tools that do not work or do not work with others' tools into the catalog.

Davis and Anderson (2003, p. 64) note:

> Standards are almost always controversial and can either be constructive and enabling or seriously counterproductive. However controversial they may be, however, some standards are essential in activities such as assuring the future interoperability of U.S. military forces or assuring reasonable degrees of composability in DoD-sponsored military simulations.

Mitigating the Risks of Poor Standards

To get standards setting right, the prototype executor will need to be careful and thoughtful in adopting or selecting standards and compliance testing for standards. To the extent possible, the prototype executor should seek expert advice from persons in industry and academia with experience with standards-setting. The implementation plan in the next chapter recommends that standards-setting authority reside in the prototype executor with the "catalog conductor," the same component responsible for the maintenance of the catalog and the investment of venture-capital-like seed money.[1] The implementation plan also calls for the creation of an advisory board comprising government, industry, and academic experts. This advisory panel should, among other things, furnish expertise and advice with regard to standards.

How to Legally and Effectively Invest in a Venture-Capital-Like Fashion

One key element of the Alt#4 business model is the stimulation of innovation or competition through the injection of seed money into the tool market in a "venture-capital-like" fashion. This would be straightforward in the private sector (corporations can invest venture capital with few restrictions), but it is problematic within the DoD. Venture capital investment directly by the DoD is not permitted; numerous federal regulations would be violated if the DoD owned an equity stake in a private firm. One answer to this problem would be following the route of the CIA and its In-Q-Tel venture capital corporation (discussed in Chapter Three). However, the Alt#4 prototype is too small to stand up an independent not-for-profit venture capital corporation as part of its limited application. Also, it is not clear that

[1] We use the term "catalog conductor" to describe the DoD component under Alt#4 charged with disbursing seed money, establishing or adopting compliance standards, and arranging for certification testing. The roles and responsibilities of the catalog conductor are discussed in Chapter Six. For those familiar with earlier presentations of the Alt#4 model, "catalog conductor" replaces "market maker" for reasons also discussed in Chapter Six.

the way In-Q-Tel serves the CIA's broad goals for possibly applicable information technology would be suitable to the more strategic and specific seed money investments called for by Alt#4. With the prototype in mind, we have sought to identify as many ways as possible for a DoD entity to disburse seed money in a venture-capital-like fashion.

As discussed in several preceding chapters, Alt#4's goal in investing seed money is not to profit but to bring new tools to the catalog in a strategic way. The DoD need not own anything at the end of this transaction, which is almost as much of a challenge as avoiding owning an equity stake. Acquisition contracts—the vehicles most familiar in the acquisition community (which generally supports procurement of training and simulation tools)—require that the government (DoD) acquire some "consideration" as part of a contract. Usually the DoD's consideration is in terms of goods or services, and the contractor's consideration is in terms of dollars.

A contracting specialist we spoke with explained to us that agreements through which the DoD spends or gives away money run on a continuum between contracts and grants. At the contract end of the spectrum, clear and direct benefit (consideration[2]) must accrue to the DoD; at the grants end of the spectrum, direct benefit accrues entirely to the awardee, and the DoD seeks only to benefit indirectly. At the contracts end of the spectrum, there are vehicles such as the cost-plus and FFP contracts discussed throughout the document, as well as cost-sharing contracts and other contract forms. All of these would be written by a contracting officer. At the grants end of the spectrum, there are various forms of grants and cooperative agreements; these would be handled by a grants and cooperative agreements specialist (different from the more common contracting officer). In between, in a sort of no man's land, are "other transaction agreements" (OTAs), which could be more like grants or more like cooperative agreements, depending on how they are written.

[2] "Consideration" is what the parties to a contract exchange, in DoD contracting parlance. Traditionally, the DoD's consideration is expressed in goods or services, and the contractors consideration is in cash value from the DoD.

Figure 5.1
The Contracts-Grants Continuum

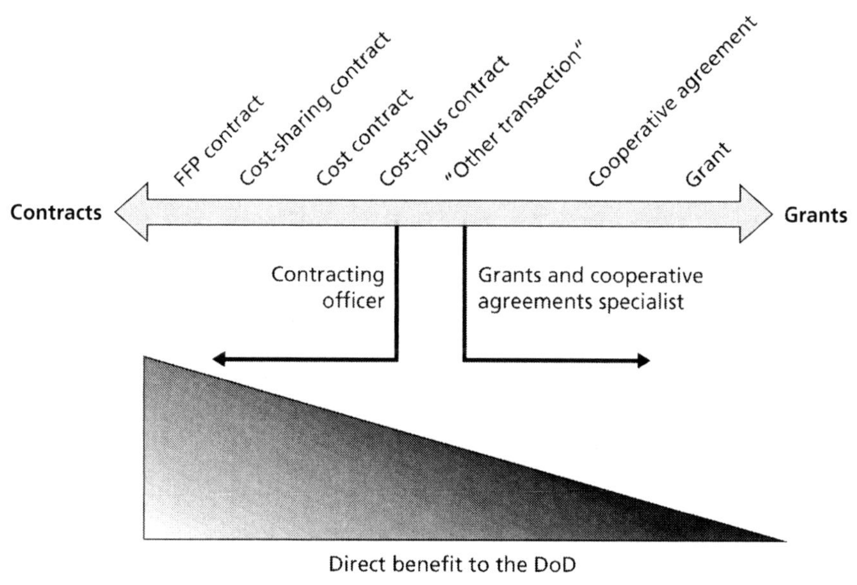

RAND *MG442-5.1*

Approaches to the Investment of Seed Money

On its face, what the Alt#4 model proposes that the DoD do is more like a grant; benefit to the DoD is indirect (making better tools available to TSPs means that TSPs provide better training to the DoD), and success of investment need not be 100 percent (it is acceptable to invest in a promising technology that does not pan out).[3]

More traditional acquisition contracts, if used in an innovative way, might also be useful vehicles for disbursing seed money. The key to writing an effective and legal contract for a seed money investment relies on contracting for a nontraditional "consideration." Rather than the DoD's consideration being delivery of a good or service, or rights to intellectual property, contracts might be written that would

---

[3] DoD guidance regarding grants and cooperative agreements can be found on the Defense Grant and Agreement Regulatory System Web site at http://alpha.lmi.org/dodgars/index.htm.

allow the DoD's consideration to be "a new tool successfully certified into the tool catalog" or some similar outcome that does not result in the DoD owning anything because of the transaction.[4]

Finally, there is the OTA. Held et al. (2002, pp. 37–38) detail the OTA:

> Recognizing the limitations inherent in its traditional contracting tools, DoD has gained a number of new contracting tools designed specifically to access the commercial technology R&D base. Unfortunately, their success in attracting NTMSs has been limited. The most important of the new tools is the Other Transaction (OT), codified in 10 U.S.C. 2371. The statutory language authorizing this contracting method provides a great deal of flexibility, because it defines OTs in the negative by stating that they are not contracts, grants, or cooperative agreements (CAs). The practical result of this negative definition is that the regulations governing the traditional contracting tools do not apply to OTs. Intellectual property rights, government oversight, cost-sharing, and business arrangements are all negotiable. In fact, by a plain language reading, it would seem that OT legislation allows any kind of agreement to conduct research between the government and a contractor, provided the agreement is in the government's interest. Thus, it would seem that the government should be routinely able to establish "business-like" arrangements with commercial businesses for research collaboration through the use of OTs.

One particularly form of OT, the Technology Investment Agreement (TIA), is another transaction cooperative agreement. It allows the DoD to share up to half the costs of applied or basic research, which, according to the guidance, appears to be applicable to the kind of seed money investments Alt#4 would like to have the DoD making.[5] As with all of these vehicles, we recommend appropri-

[4] We say "might be able to be written," because we did not consult with a contracts attorney regarding this proposal and are relying on a plain language interpretation of the FAR. All actual innovative efforts to disburse seed money should be made in consultation with a DoD contracts lawyer.

[5] See http://alpha.lmi.org/dodgars/tias/tias.htm.

ate consultation with a contract lawyer and contracting or grants and cooperative agreements experts before using a TIA for Alt#4 efforts.

Venture-Capital-Like Activities Beyond the Money

Investing in a venture-capital-like way goes beyond just spending money; how and where venture capital firms choose to invest is also an important part of the process. Held et al. (2002, pp. 41–42) note:

> Once funding has been raised, the management of a venture capital fund goes about the process of evaluating investment opportunities and selecting companies for funding. A primary function of the venture capitalist is to gather information about potential markets, technical feasibility, competition, and other facts that will impact the probability that a new business will succeed. This knowledge comes from a number of sources, including past experience, contacts in the market segment, other venture capitalists, trade journals, and the business plans submitted by entrepreneurs looking for funding. Using a combination of experience, analysis, advice, and intuition the venture capitalist decides which ventures to fund and the extent to which they will be funded. As mentioned above, venture capital involves various funding mechanisms, though equity financing is the most common. In general, venture capitalists fund relatively new and rapidly growing companies. There are a couple of reasons for this. First, newer companies tend to be more efficient in that they have much less overhead and a core staff more directly affected by the success of the company. Second, and perhaps more important, new and growing companies have a greater potential for the high rates of return that venture capitalists require.

Fortunately, an Alt#4-focused seed money investment is not concerned with profit but should be concerned with the ability of seed money recipients to bring their product to market and the potential applicability of those products to emerging training needs.

Identifying Emerging Needs

The Alt#4 business model relies on the participants knowing about emerging training needs. The training users must identify their emerging needs so that they can share them with the component responsible for seed money investment and write solicitations and contracts for training support to meet those needs. The catalog conductor must identify emerging training needs so that they can encourage tool vendors to create tools to meet these needs, with seed money investment if necessary. The TSPs need to know about emerging training needs so that they can prepare to compete to meet those needs and notify the tool vendors about what tools they anticipate needing. The tool vendors need to know about emerging needs so that they can make tools to satisfy them. This scheme is well aligned, in that everyone needs to know about emerging training needs and is best served by sharing knowledge of those needs with other model participants. However, just because all model participants need to know about emerging training needs does not make those needs any easier to identify. Further, to get effective FFP contracts, training users must be prepared to specify their needs very precisely (as discussed in detail in the next two sections). Needing to specify needs precisely at the point of solicitation and contract can be separated from anticipating and circulating broad contours of emerging needs.

Ultimately, the system implicit in Alt#4 is no worse than the present system for identifying emerging training needs and the technology needs required to meet them. In fact, given that private industry is recognized as being much more agile at developing and procuring new tools than the government is (Held et al., 2002, p. xv), once pending needs are identified and passed around to various model participants, private sector tool vendors should make tools to meet those needs available for use more quickly than would be the case if they were to be the result of a government procurement.

Soliciting Requirements Versus "Desirements"

The DoD has already made great strides toward eliminating "onerous, complex, [and] costly" defense-unique military specifications (milspecs) from its acquisition processes (Schmidt, 2000, p. 10). The DoD's move toward specifying performance requirements, rather than "how to" specifications for contractors, represents great forward progress. The next critical step on this logical trajectory is to a solicitation and contracting system that makes clear which product or service specifications are actually requirements (*must* be part of what is delivered) and which are "desirements" (things that are desired but might be abandoned for cost savings, speed of delivery, or other negotiated efficiencies).

Carney and Oberndorf (1997, p. 4) note that "there is widespread agreement throughout the software community on the importance of a requirements specification; anecdotal evidence suggests that the requirements specification can be the single most important factor in the success of an acquisition." This becomes particularly true when using FFP contracts, since a change in a requirement that might simply increase cost and delay delivery under a cost-plus contract will likely meet fierce resistance from an FFP contractor.

Failing to adequately specify requirements and overconstraining the possible solutions through restrictively expressed requirements are two edges of a sword that both pose risks to contracts under the Alt#4 business model.

Solicitation Solutions

Alt#4 is predicated on a desire to obtain innovative, high-quality, cost-effective training for the DoD. Implicit in the TC AoA process was a desire to be able to buy the "90 percent solution" to a training problem, quickly and inexpensively. To be able to do so, training user solicitations must balance between inadequately specified and overspecified requirements.

The solution is to clearly separate "requirements" from "desirements" and, where possible, to solicit reasonable solutions to clearly defined problems rather than specific answers to those problems. Solicitations should have language that makes clear what is desired in terms of "training targets" or "goals." Goals should also be prioritized, if possible. If a training user is open to a cheaper "90 percent solution," that should be indicated in solicitations, as well as the most important goals. Drezner and Leonard (2002, p. 10) note the success realized during the acquisition of Global Hawk: "Performance parameters were stated as goals rather than as requirements, allowing for a high degree of design flexibility."

Performance Measurement in Firm-Fixed-Price Contracts

The Alt#4 business model's central precept is that the DoD buys training with firm-fixed-price contracts. FFP contracts are not without risks, however. Uncertainty-based risk is not transferred to the DoD (as it is in cost basis contracts), so vendors are likely to include wide "safety" margins in their bids beyond what they think their service will cost (see the more extensive discussion under "risk" below). Still, this inflated cost will be known at the start (unlike with a cost-plus contract), and the diminution of uncertainty with repeated successful deliveries coupled with competition in the TSP market creates incentives for reduced bids and further savings for the DoD.

Similarly, the same incentive that pushes TSPs to reduce costs also pushes them to reduce service to the minimum acceptable level. Mitigation of this risk requires very careful contract-writing. If the training outcome to be met has firmly established exit and evaluation criteria, and the contract indicates that the TSP is paid only for each trainee who attains a certain measured level of competence, then the TSP's incentives will align toward maximizing trainee attainment of the target standard at a minimum cost. If the training success metrics are well constructed and correspond to desired training outcomes, then everyone will get what they want. If, on the other hand, the contractually specified success metrics are *not* good measures of desired training outcomes, TSPs have no incentive to deliver desired training outcomes and the DoD is left with limited contractual recourse.

Clearly expressing requirements and identifying performance measures is easier for some classes of skills than for others. Training tasks that include recognition and procedural skills, such as friend-or-foe identification or accurate calls for fire, should be relatively easy to define performance measures for. Classes of skills that require tasks that focus on complex cognitive skills or judgment, however, are much harder to specify training requirements and performance measures for. How, for example, do you contractually measure a trainee's completion of a requirement that reads: "creates 4 well-designed, adaptive courses of action, decides on the best, and provides aggressive, adaptive C4ISR for the mission?" Where performance criteria cannot be specified, the risk associated with FFP contracts increases dramatically. Alt#4 is not appropriate where performance criteria cannot be specified.

To minimize performance risks from FFP contracts, training users need to make sure that they develop clear measures associated with their target training outcomes. Failure to correctly match measures with desired outcomes could result in the general phenomenon of "teaching to the test," where TSPs would instruct trainees in the skills necessary to satisfy the performance measure and *not* the underlying training need (Klerman, 2005; Hamilton, Stecher, and Klein, 2002; Stecher and Barron, 1999). Caution, precision, and forethought are the only risk-mitigation strategies RAND has identified.[6]

Risk

One risk the prototype faces has to do with the transference of risk itself inherent in the Alt#4 business model. Under a cost-plus contract, the DoD bears the financial risks related to uncertainty, whereas the contractor avoids these risks by being paid for whatever

[6] Modifying the model would be another strategy to mitigate risk. For example, starting with cost-plus contracts and then evolving to firm-fixed-price contracts could mitigate risk but would yield no useful evaluation information in the limited life of the test.

costs it incurs. In principle, cost-plus contracts allow the DoD to avoid output risk. That is, by agreeing to buy an output whatever the cost, the DoD minimizes the risk that the contractor will go out of business or otherwise fail to provide the output. In practice, however, cost-plus contracts often result in astronomical costs *and* output failure (JSIMS, most notably).

Under Alt#4, the DoD transfers uncertainty-based risk to the TSPs through firm-fixed-price contracts. These service providers will mitigate risk by using standard commercial practices; they will estimate risk and increase their prices accordingly. However, the DoD retains the output risk. If the TSP fails to deliver the contracted training, the DoD avoids much of the fiscal cost of that failure (depending on the exact terms of the contract) but does not receive the training support contracted for.

For the Alt#4 prototype, there is a nonnegligible risk of failure that includes the failure of a TSP to deliver the contracted training output. To mitigate this risk, the prototype contracting/grants support should assist the training user in assessing the bids received from the TSPs and evaluate these bids for best value and likelihood of successful delivery.

Market Risk

The creators of Alt#4 made several implicit assumptions about the existence, robustness, and willingness to participate of the two commercial markets: the training service providers and the tool vendors. There is a risk that these assumptions will not bear out in practice. Alt#4's creators recognized an existing training service provider community with numerous participants variously providing live, virtual, or constructive training to the DoD and other government and law enforcement agencies on a contract basis. Alt#4's creators also saw a burgeoning community of game developers and modeling and simulation designers, some of whom do business with the DoD, some of whom want to do business with the DoD, and some of whom want nothing to do with the DoD. They assumed that this first market (TSPs) was sufficiently capable and engaged with the DoD to provide competitive responses to Alt#4 compliant solicitations by training users. They further assumed that the tool vendor market would be willing and able to produce tools for the catalog but that innovation and competition in this market might not initially be sufficiently robust without additional stimulation (hence the seed money activities associated with the model). RAND has seen no evidence to suggest that these observations and assumptions are incorrect, but if they do prove to be incorrect, there is a risk the prototype will fail.

Unfair Competition from Government Furnished Equipment

One concern raised by members of the "Macrosystems" team that created the Alt#4 business model was the possibility of "unfair" competition from government furnished equipment (GFE) provided essentially free to TSPs.[7] Although competition from GFE may be a challenge to tool vendors that wish to compete directly with GFE simulation tools, we remain unconvinced that providing existing GFE free to TSPs impedes the DoD's ability to acquire best value training. To compete with GFE, tool vendors must provide a tool that delivers better value than that of GFE; that is, new tools must be sufficiently better than existing GFE to justify their price tag based on a best value assessment. When and if new tools cross that relative best value threshold, they should expect to replace GFE in training delivery.[8]

---

[7] Traditionally, DoD program offices for DoD-owned simulation tools have been keen to provide these tools to TSPs providing training to the DoD. This is, after all, what the tools were acquired for, and many program offices have use of their system as one of their performance metrics. Program offices provide these simulation tools as GFE, either without cost or with an implicit trade in kind agreement ("you can use our simulation tool, if you add functionality that will allow it to do 'X'"). If already-owned DoD simulation tools enter the catalog as GFE with an effective price tag of "free," commercial tool vendors face stiff competition even if their simulation tools are considerably better than the GFE tools.

[8] We remain unconvinced that free GFE actually threatens the Alt#4 model's ability to function, but we have identified several possible solutions should it prove to be a problem.

If the problem is GFE being free, why not associate a price with it? We recognize two problems with attempting to set a price for GFE: how to set the price, and the fact that the DoD can not legally "sell" its tools to TSPs. One solution attempts to solve these underlying problems. Contracts in other acquisition areas (especially those that require minerals or precious metals as part of the contractual good) are often bid with and without specific GFE; the contracting entity then makes its best value bid assessment including comparison of the

Risk That Prototype Will Not Really Be Alt#4

Chapter Three summarizes seven business models for the acquisition of simulation training tools and training support. Several of these models have elements in common with the Alt#4 business model. As in all bureaucracies, DoD personnel are most comfortable doing what they are familiar with. Even when "innovation" is a stated goal, there is a tendency to do "new" things in a familiar way.

The risk of the prototype not actually implementing the Alt#4 business model is compounded by the complexity of the model and the different interpretations of the model held by different stakeholders. The RAND team spent a considerable amount of time at the beginning of the project trying to understand the Alt#4 business model and to track down and understand the intentions of its creators and its core logic. During that process, we received many in-

---

two bids from the same vendor. This allows the DoD to realize savings from providing GFE without actually "selling" or receiving payment for it (and thus violating a host of federal statutes). This works best for situations where the GFE in question is a commodity for which there is a preestablished market price (again, precious metals is a good example). However, this approach suggests solutions to both problems with providing free GFE. First, TSPs that wish to consider using GFE can be asked to bid with and without the GFE; the difference between the two contracts (adjusted for the cost of non-GFE tools that will fill the same modular hole) can be used as a real market price to determine the "cost" of GFE. Once that "cost" is determined (either in a single instance or averaged over multiple "with and without" bids from different contractors), it can be added to the cost of "with GFE" bids for the purpose of evaluating proposals. In other words, cost savings from GFE being actually "free" from the DoD standpoint can be concealed from those evaluating the bids for the best value contract so that GFE seems to have a cost from the perspective of bid evaluation.

Another solution would be to refuse catalog entry to DoD-owned simulations. Ultimately, the Alt#4 business model would keep the DoD out of the business of owning simulations, and preventing DoD-owned simulations from being used by TSPs should hasten the transition.

Another solution would be to privatize all existing DoD-owned simulation tools. "GFE" could be given to tool vendors (sensibly, the contractor who built each simulation in the first place). These vendors would then own the tools, would set market prices appropriate to them, and could legally sell them in a way that the DoD cannot. Of course, this may not be a practical solution until or unless the DoD decides to get out of the business of owning simulation tools.

A final solution would be to make all DoD-owned simulations, software and hardware, "open source." This would allow TSPs and tool vendors to examine the underlying design and implementation of GFE tools and decide if they should build on the existing GFE, use their own, or build something entirely new.

dividuals' interpretations of the model; all shared some fundamental agreement, but certain interpretations had "drifted" from the original concept and were not consonant with some of Alt#4's original core goals. For example, as recently as early September 2005, we saw draft briefing slides suggesting that government labs might receive some of the seed money investment as part of the prototype—an intention that is clearly not consonant with Alt#4's goal of getting the DoD out of the business of owning simulation tools.

Alt#4 as presented in Chapter Two of this report is true to the original logics and incentives imagined by the model's creators during business game #2. Any modifications we made from the original implementation concept were to ensure that the model as a prototype would remain true to the underlying logic of the original Alt#4 proposal. This is the model the DoD wishes to test with the prototype. Given the DoD's desire to evaluate the Alt#4 business model, it would be unfortunate if the prototype "succeeds" at delivering quality training at reasonable rates but is not, in fact, a prototype of Alt#4 but is rather a hybrid business model; one would then never know if Alt#4 were actually a better model.

CHAPTER SIX

Critical Elements of a Prototype Implementation Plan

Taken together, Chapters Three, Four, and Five suggest that the Alt#4 business model clearly has benefits to offer the DoD but that attempting to realize these benefits is not without risk. This chapter lays out the critical elements of a plan to implement a prototype of Alt#4. The RAND team sought to frame a plan so that a prototype of the Alt#4 business model implemented following this guidance will

- be able to function legally within the DoD context;
- be true to the model as envisioned by the TC AoA business game team that conceived it;
- adhere to the model principles validated by economic theory in Chapter Four; and
- be well positioned to implement mitigation strategies against the risks identified in Chapter Five.

To develop this plan, the RAND research team relied on information from a variety of sources: The TC AoA final report; the outbrief prepared by the "Macrosystems" team after TC AoA business game #2, interviews with business game #2 participants, discussions with project sponsors, interviews with Joint Forces Command (JFCOM) personnel charged with prototype execution, interviews with SOCOM personnel with knowledge regarding Joint Close Air Support (JCAS) and Alt#4, review of literature/theory from economics, and case studies of similar efforts in defense or in a broader business context.

The chapter is organized as follows: First, we present criteria for the selection of prototype learning objectives and executors. Then, we identify the minimum set of organizational components or entities to realize the business model. This is followed by a list of the roles and activities that must be assigned to those components, with recommendations on which components should conduct which activities. The chapter concludes with a discussion of the order in which prototype activities should begin and an implementation checklist that summarizes recommended steps to implementation.

Choosing a Learning Objective and a Prototype Executor

During the performance period of the research effort that produced this report, OSD chose a learning objective and executor for the prototype.[1] As of this writing, our understanding is that JFCOM will execute the prototype with JCAS as the learning objective.[2] The prototype effort is slated to be funded for a total of $15 million over three years, beginning in FY 2006.

Even though JFCOM has been selected to test the prototype Alt#4 in the area of JCAS, we include criteria for selecting the prototype executor and learning objective for four reasons: First, since we shared a list of selection criteria with the sponsor in draft form before the final selection of an executor and learning objective, we include these criteria for completeness. Second, since we expect the final draft

---

[1] By "learning objective," we mean to denote the content area or targeted training tasks that need to be addressed by the prototype. Referring to the targeted training need as a "learning objective" is an artifact of early participation in the prototype planning process by an individual at SOCOM. Since numerous stakeholders have become familiar with the "learning objective" nomenclature (also used in the RAND briefing that informs this section), we retain the term.

[2] The GAO (see U.S. Government Accountability Office, 2003, for example) has found that readiness for air support of ground forces still requires improvement. Our interviews suggest that significant gaps remain in existing JCAS training capabilities.

of this document to be made available to the JFCOM program executors, we hope that they can benefit by seeing which executor selection criteria play to their strengths and which suggest they will need to develop additional strengths. Third, some of the criteria for choosing a learning objective may still be useful because the prototype executors choose specific tasks or exercises within the broader functional area of JCAS for which to purchase training support. Finally, if the prototype succeeds and is followed by a DoD effort to expand the application of the business model, these selection criteria might suggest additional learning objectives to add incrementally as the program expands.

Our choice of selection criteria was based on our understanding of the requirements of the Alt#4 model and the realities of the DoD training and acquisition environment.

We recommend that the learning objective on which the prototype will focus satisfy as many as possible of the following criteria:[3]

1. The learning objective should be something for which it will be easy to stimulate and engage both the TSP marketplace *and* the tool vendor marketplace.
2. The learning objective should lead to well-articulated training requirements that are not overspecified (i.e., would allow more than one possible solution) so that they can easily be written into solicitations attractive to TSPs.
3. The learning objective should contain training requirements that are met with discrete, well-defined events or learning outcomes with known or easily derived performance standards/metrics. This should be something that will be easy to get FFP bids on, with TSPs, the DoD, and evaluators easily able to determine whether the training provided is meeting requirements.
4. The learning objective should include training requirements that can be met through many training events or frequent repetition of a few events. If the learning objective is a high activity area, it

[3] These criteria were transmitted to the project sponsors in draft form on July 27, 2005.

will yield multiple observation points, the opportunity to observe change over the course of the prototype, and potentially multiple contracts to be let over the course of the prototype.

5. The learning objective should include training requirements that need some new technology or new integration of technology to meet but will be feasible in the short-term. It could be replacing outdated simulations (increase technology) or improving on low-fidelity simulations (increase fidelity) or tasks not simulated (increase availability). What those technology needs are must be clearly expressible. Clear, reachable technology goals will be easiest for the DoD to invest in or stimulate.
6. Technology needed to satisfy training needs for the learning objective should *not* be entirely (or even mostly) met by existing government off-the-shelf simulations. To demonstrate the entire Alt#4 business model, the prototype must be able to stimulate the tool vendor marketplace, not just get TSPs, to bid FFP contracts to use off-the-shelf simulations.
7. Training technology needed to satisfy the learning objective should lean toward the "lightweight" end of the simulations spectrum so that needs can be met reasonably quickly and potentially in a number of different ways.
8. The broader learning objective should include multiple unmet needs (in both technology and training) that can be prioritized and not solved immediately with a single solution. This will allow the prototype to show multiple (small) instances of success and some of its iterative character over the duration of the test period.
9. The learning objective should have training users who are willing to participate. It is critical that they be willing to bid their training with FFP contracts that comply with the requirements of the business model.
10. Funds should already be budgeted to satisfy the training requirements encompassed by the learning objective. The prototype budget alone is not large enough to buy the training and fund the prototype administration and seed money investments.

11. The learning objective should be, or be part of, one or more recognized Joint Training Requirements. Being joint should help avoid inappropriate parochial input on either the training or the technology side; approved joint requirements bring a necessary level of legitimacy to the training task side of the effort.
12. Multiple learning objectives should at least minimally satisfy these criteria. The chosen learning objective should be the best of many, not the "least worst" or only choice.
13. The learning objective selected should be plausibly generalizeable ("If it worked in that training area, it will work in others, too") and have face validity (should make sense as an example).

We recommend that the selected host or prototype executor meet as many of these criteria as possible:

1. The prospective host must be interested and engaged (the host should want this).
2. The prospective host must be flexible, able to either conduct catalog conductor activities itself or liaise effectively with an external nonprofit venture capital firm (depending on which implementation proves most feasible).
3. The prospective host must be of sufficient size and staffing to execute the business model.
4. The prospective host should have experience with both technology and training procurement. The executor should have access to contracting personnel familiar with science and technology development, systems acquisitions, and training, as the business model involves action on training procurement and technology investment.
5. Ideally, the prospective host should be a training user for the learning objective. Executor motivation to make the business model work is likely to be higher if they are "doing something for themselves."
6. The prospective host should be similar to other organizations to which the business model may eventually be expanded, so that

prototype results are plausibly generalizeable ("If it worked for them, it will work for others, too.")

Components/Entities and Their Responsibilities

In the discussion of economic theory and the incentive structure of the Alt#4 business model in the chapters above, the prime actor is the DoD. However, an effort to realize a prototype requires greater specificity with regard to who must do what. Even with the decision that JFCOM will execute the prototype, implementation planning is not complete simply by substituting "JFCOM" for "DoD" in the previous chapters of this report. This section details our recommendations for the components or entities that JFCOM should create or assign within itself to execute the prototype. These proposed components are based on components identified in the original Alt#4 outbrief and on an understanding of general business practice and DoD organizational requirements. Following each component, we identify and tentatively assign roles and responsibilities to each. Exigencies of execution may cause the prototype executor to need to assign some roles and responsibilities to slightly different combinations of components. We recommend that all roles and activities listed below, regardless of which component conducts them, be assigned to one or more components and included in the prototype implementation effort.

Figure 6.1 summarizes the core role of each component in the business model.

Governance/Oversight

Governance and oversight will play an important role in the prototype implementation. Because the Alt#4 business model is innovative, the governance component has responsibility for making sure that the other components adhere to the requirements of the business model and generating and approving course corrections as unanticipated challenges emerge for the prototype. These roles are in addition to the more conventional (and still important) oversight role. A more

Figure 6.1
Central Roles of the Various Components of the Alt#4 Business Model in Relation to One Another

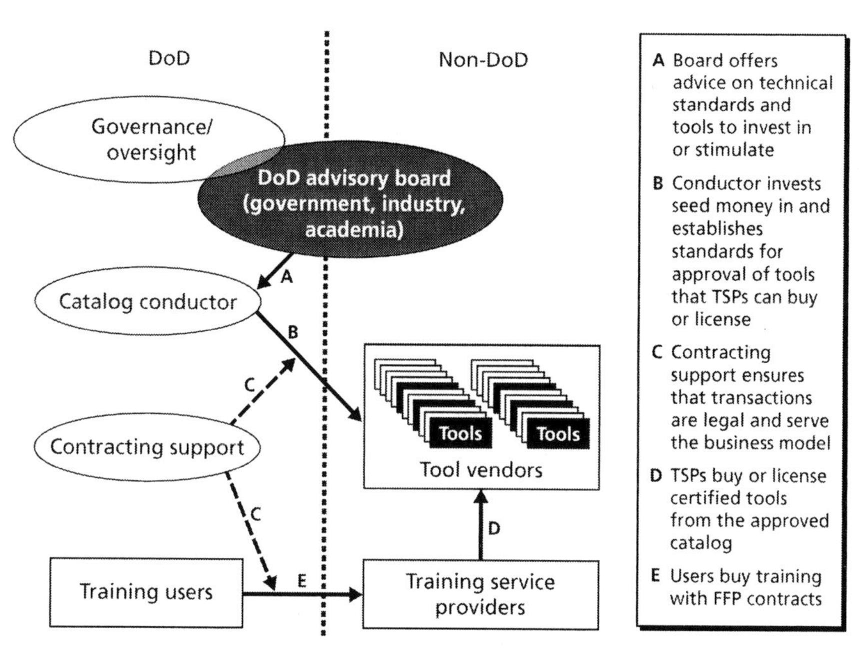

RAND *MG442-6.1*

extensive list of roles and responsibilities that we recommend be assigned to the governance component appears below.

The number and level of governance and oversight entities will be partially determined by DoD organization and regulations. JFCOM may wish to establish an oversight and management entity at the level of the prototype executor to report to higher-level oversight and be more involved with the moving parts of the prototype implementation; in that case, some of the roles and responsibilities discussed below may shift to this additional entity; regardless, the core tasks included in the roles and responsibilities remain the same.

Governance Roles and Responsibilities

- manage;
- provide oversight;
- develop and enforce policy;
- ensure compliance among participants;
- establish and track performance measures;
- evaluate success of prototype, consider future expansion;
- educate and report prototype outcomes and progress to services, other stakeholders, and higher DoD authorities;
- review and revise business model as needed; and
- allocate budget and resources (determine funding available for seed money investment).

The Catalog Conductor

The Alt#4 model as originally conceived during the TC AoA had a component called the "market maker." Envisioned as a consortium of government, academia, and industry personnel, market maker responsibilities focused on the tool vendor marketplace and included establishing standards and compliance testing, investing seed money in a venture-capital-like fashion, and collecting and relaying emerging training user training needs to the tool vendor market. Although market maker is certainly evocative language and suggests an entity that will ensure that the tool vendor market is ready to serve the tool needs of the TSPs, the term already means something else. Market makers are an important part of over-the-counter stock exchanges and are firms that stand "ready to buy and sell a particular stock on a regular and continuous basis at a publicly quoted price" (U.S. Securities and Exchange Commission, 2000).

Since the term already has a clear technical definition that does not correspond with the activities of the entity bearing that name in the original Alt#4 conception, we urge all participants and stakeholders to discontinue its use. We propose "catalog conductor" as a term that has not been used in other contexts and effectively captures the core goal of the component.

The catalog conductor is responsible for filling the catalog with tools useful to satisfying training needs. This entails three key tasks:

1. discovering existing and emerging training needs and publicizing those needs to the tool vendor market;
2. establishing or adopting technical standards and compliance testing for tools to be included in the catalog; and
3. ensuring that the catalog is populated with a variety of competitive tools.

The catalog conductor will accomplish this through interactions with tool vendors and use of seed money to encourage the development of innovative or competitive tools as previously discussed.[4]

Catalog Conductor Roles and Responsibilities

- identification of emerging needs and requirements;[5]
- tool market research (facilitated by the advisory board);
- tool market liaison (facilitated by the advisory board)—advertise emerging training needs;

[4] Potential methods for this were discussed in Chapter Five. The advisory board, discussed next, plays an important role with respect to criteria for, magnitude of, and means to make such allocations.

[5] For the tool market and the training support market to deliver responsive tools and training, they need as much advance warning as possible about emerging needs. New tools cannot be created overnight, and the TSPs cannot deliver training with tools that do not yet exist. This problem will challenge any effort to meet emerging needs; barring prescience or luck, it is difficult to know what to build before it is needed.

In the current implementation, every component, even those outside the DoD, has an interest in the identification and sharing of emergent needs. The two DoD-side entities with responsibility for collecting and acting on those emergent needs are the catalog conductor and the training users. Presumably, the training users already have a process in place for determining their needs and requirements. The catalog conductor should be in regular contact with those responsible for generating training user needs. To the extent possible, the catalog conductor should try to "get out in front" of these emerging needs, trying to recognize "proto-needs" before they become formal training requirements and conducting market research or stimulation to make sure that the tool market will be ready to meet that need when it fully emerges.

- establishment or adoption of standards for entry into the catalog;
- periodic review or revision of standards as needed;
- maintenance of a catalog list (catalog entries themselves are still proprietary property of the vendors that made them);
- identification and prioritization of training needs and the technologies likely to meet them (with guidance and support from the advisory board and governance);
- production of solicitations for development of tools in emerging areas of need or areas with limited competition;
- strategic injection of seed money into the tool market;
- compliance certification;[6] and
- leveraging of other funding sources for investment in targeted technology areas (small business innovation research [SBIRs], joint and service R&D funding, the Defense Advanced Projects Agency [DARPA], etc.; governance may be able to help with this, too).

Advisory Board

As noted above, the developers of Alt#4 originally conceived the component now referred to as the catalog conductor as being a consortium of industry, academic, and government representatives. Legally, the DoD cannot have such a consortium making decisions or disbursing funds in the way the catalog conductor must.[7] However, in the original conception, this consortium would also leverage its

---

[6] Note that existing JFCOM verification, validation, and accreditation (VV&A) capability may be able to do certification testing if it has the ability to test the standards the catalog conductor adopts. Standards should be decided on first, then the decision to do or buy compliance-testing should be made. When making compliance-testing decisions, not only capability to do the testing but throughput and cost must be considered. Cost to tool vendors should be as low as possible (to give incentives to list tools), and throughput capacity should be sufficiently high for vendors who want to get their tools tested to do so without lengthy delays.

[7] We examined guidance regarding Consortium Member Agreements and concluded that such arrangements were suitable neither to the activities of the catalog conductor nor the advisory board.

experience, technology knowledge, and industry contacts to facilitate the smooth functioning of the Alt#4 model.

To avoid the loss of this valuable advice and input from the business model, we recommend that the DoD establish an industry, academia, government consortium as a Department of Defense Federal Advisory Committee, in compliance with Title 41 of the Code of Federal Regulations, Subpart 102-3, "Federal Advisory Committee Management," and DoD Directive 5105.4, "Department of Defense Federal Advisory Committee Management Program," February 10, 2003. This advisory board should be chaired by an involved member of the governance/oversight component, both to provide direct representation between the two components and to comply with regulations requiring a designated "officer or employee of the Federal Government" to chair or attend all advisory committee meetings (CFR, Title 41, Subpart 102-3, section 10e).

Regarding the composition of this advisory group, we recommend a broadly representative group not unlike the groups recruited for the TC AoA business games. Those recruiting for the advisory group should consider placing inquiries with existing organizations and consortia, such as the Network Centric Operations Industry Consortium, the National Modeling Analysis Simulation and Training Coalition, the National Training Systems Association, the Simulation Interoperability Standards Organization, or the Government, Academic, Military, Entertainment and Simulation (G.A.M.E.S.) Synergy Summit.[8] Recruiters should also contact universities with prominent modeling and simulation groups, such as the University of Southern California, Purdue University, or Carnegie Mellon University.

In addition to representation from the advisory board, government representatives should be sought from agencies or components with relevant experience, such as the Defense Modeling and Simulation Office, DARPA, and perhaps even individuals from the Depart-

---

[8] See the respective Web sites at www.ncoic.org, www.nmastic.com, www.trainingsystems.org; www.sisostds.org, and www.synergysummit.com.

ment of Commerce with experience relevant to tied markets and the effect of subsidies on competition.

Held et al. (2002, p. 50) suggest that it is important that DoD venture-capital-like efforts be supported by a staff that contains a "mix of personnel with business, technology, and government experience." The advisory board needs to fill relevant gaps in the experience and expertise of the catalog conductor staff.

Organizers should target an advisory group size that is manageable but will still provide a sufficient mass of experience and advice when the entire group cannot convene. Ideally, periodic meetings of the full group should be able to draw between eight and 12 attendees, but much of the business of the advisory group can be conducted with potential participation from the whole board via email or another electronic discussion forum.

Advisory Board Roles and Responsibilities

- be responsive to catalog conductor advisory needs;
- provide important insights from industry experience;
- advise catalog conductor on standards;
- advise catalog conductor on tool market and tools to consider for seed investment;
- assist catalog conductor with market analysis/research;
- leverage private sector contacts to encourage participation in tool vendor market/catalog listing; and
- leverage private sector contacts to provide feedback to catalog conductor and governance regarding what is and is not working on the private sector side of the business model.

Contracting/Grants Support

To function, the Alt#4 business model requires two types of carefully written agreements: the FFP contracts between the training users and the TSPs and the agreements (be they contracts, grants, cooperative agreements, or something else) associated with the catalog conductor's seed money investments. Although the seed money investment agreements may become routine once the appropriate vehicle is discovered, training users are likely to require ongoing support preparing FFP contracts, given the need to carefully specify observable training completion criteria for each one.

To write these agreements and to provide the necessary support to the training users and the catalog conductor, we recommend that the prototype implementation include a contracting and grants support entity with access to a contracting officer and a grants and cooperative agreements officer. This either can be through a new office dedicated to the prototype effort or could be an additional effort for an existing contracting office and grants and cooperative agreements office. This latter choice is likely more practical. If relying on existing offices, the prototype executor must take care that the contracting and grants and cooperative agreements officers involved are willing and able to support the kind of innovative contracting required by the Alt#4 business model. As one participant of the JFCOM Innovative Acquisition Strategy Offsite (August 3, 2005) correctly noted: The prototype needs contracting personnel "who don't know how it can't be done."

Note that because the appropriate vehicle with which to make seed money investments may fall under the purview of a grants and cooperative agreements officer (see the discussion in Chapter Four), the function that is assembled or selected should have access to personnel with the appropriate expertise. Because Alt#4 necessitates innovation in the contracting arena, the contract support also needs to have ready access to a contracting lawyer. Once vehicles for seed money investment are established and made routine, the need for consultation with specialist personnel should diminish.

Contracting/Grants Support Roles and Responsibilities

- write contracts and solicitations;
- help catalog conductor prepare effective, legal contracts or cooperative agreements to disburse seed money; and
- help training users write contracts that meet training user needs *and* meet requirements of the business model: FFP contracts,

conflict of interest clause between TSPs and tool vendors, and TSPs using only catalog certified tools.

Other Participants

The components listed above are the elements of Alt#4 model that the DoD must establish during the prototype implementation. Three other important types of entity that play an active role in making the Alt#4 business model work: the training users, the TSPs, and the tool vendors.

Training users are clearly part of the DoD but are not a part or component that is formed explicitly for the prototype, nor does their participation in the prototype represent a radical departure from or addition to their normal activity portfolio. Training users already contract for training support; the Alt#4 business model simply demands that they contract for training support in a specific *way*—firm-fixed-price per training outcome contracts with clauses that prevent conflict of interest between TSPs and tool vendors and that require that TSPs use tools certified in the catalog. Training users may require support from prototype entities (such as the prototype contracting/grants support) to successfully bid and execute contracts compliant with the Alt#4 business model, but the training users are more a part of the prototype audience than they are a core prototype component.

More clearly divorced from the prototype (but no less critical) are the TSPs and tool vendors. A significant number of TSPs and simulation tool vendors already exist. Getting these TSPs to bid in response to prototype-compliant training user solicitations and getting tool vendors to enter tools into the catalog are important parts of the Alt#4 implementation effort.

Training User Roles and Responsibilities

- informally share existing and emerging training needs with catalog conductor as early as possible;
- identify easily measurable standards or tests to certify satisfaction of training needs, so that FFP contracts can be clearly stated as

payment per delivery of satisfaction of that training standard; and
- solicit and buy training support on FFP contracts that comply with the other requirements of the Alt#4 business model.

Budgeting the Prototype Implementation

The prototype will have a limited budget (our current understanding is that the prototype is budgeted for $15 million over three years). These funds need to support the operation of the core prototype components (governance, catalog conductor, advisory board, and contracting/grants support), including administrative costs and the personnel costs associated with executing the prototype. The prototype budget is also the source for any seed money the catalog conductor will invest in the tool vendor market and must cover costs associated with compliance testing as well. As much as possible of the prototype budget should be reserved for catalog conductor activities.

The prototype budget *should not* be used to pay for training support. Training users already have funds with which to buy training and training support; the prototype simply asks them to do it in a new way. Training users should be convinced to participate because of the efficiencies they are likely to realize through the prototype. Expensive incentives to training user participation, such as prototype budget funds for training support, should not be required.

Regarding the size and number of seed money investments, discussions with "Macrosystems" team members revealed that they envisioned small investments, on the order of $50,000 to $500,000 each, and the number made in each iteration (for a given tool gap) would differ depending on market conditions but would generally be between three and seven. The goal would be to make sure that at least two (preferably three) functional tools emerge and compete in the catalog for each newly identified tool gap.[9] Note, however, that cata-

---

[9] One member of the "Macrosystems" team pointed out that not every catalog conductor investment needs to result in a useful tool. Although it would be wonderful if every seed

log conductor seed money need not be the only source for innovation or innovation support. A resilient tool vendor market might produce three new competing tools on its own; other government or DoD technology investment programs might contribute as well. Catalog conductor market research (with help from the advisory committee) will allow the catalog conductor to determine how much seed investment a given tool gap will require. Because of the relatively short (three-year) duration of the prototype, seed money invested during the prototype should focus on tools that can be built and entered into the catalog as quickly as possible (while still serving prototype learning objective training needs).

In What Order Should Activities Commence?

When active, the Alt#4 business model involves several cyclical and continuous processes and a great deal of simultaneity. Unfortunately, it is unrealistic to expect the prototype to emerge fully formed and functional, like Athena from the head of Zeus. It must be established one piece at a time and has to begin somewhere. This section provides a notional order of activities for the initial implementation of the prototype of the Alt#4 business model.

1. Stand up all entities/components. Priority: governance, catalog conductor, advisory board, contracting/grants support.
2. Governance identifies one or more training users.
3. Governance and catalog conductor get training users to describe new training needs, including anticipated needs over the prototype period.
4. Convene advisory committee. At first meeting, advisory committee discusses identified training needs, makes recommendations

money injection resulted in a tool entering the catalog, the target, he asserted, should be closer to 50–70 percent of seed money investments yielding tools, and perhaps a lower "success" rate for tool gaps that are very difficult to close.

about standards and recommendations about applicable technology solutions to the catalog conductor.

5. Catalog conductor adopts or establishes compliance standards.
6. Catalog conductor issues solicitations for use of seed money to develop (compliant) tools to meet the new need. Relatively short solicitation period.
7. Catalog conductor reviews proposals; perhaps directs questions about feasibility of proposals to advisory board (via electronic forum).
8. Working with contracting/grants support, catalog conductor makes initial seed money awards.
9. Catalog conductor arranges for compliance testing (VV&A).
10. Catalog conductor advertises standards and availability of compliance testing and encourages vendors to list their existing compliant tools.
11. With assistance from contracting/grants support (or from catalog conductor or even from management side of oversight), training user defines training needs as goals and key performance parameters and thinks about how achievement of those parameters can be certified/clearly demonstrated.
12. With assistance from contracting/grants support, training user solicits bids for training service support, making clear that the accepted bid will be for a contract that is FFP; requires use of catalog tools; and precludes conflict of interest (COI) between TSP and tool vendors.
13. With assistance from contracting/grants support, training user accepts best value bid and enters into contract that complies with the requirements of the Alt#4 business model.

CHAPTER SEVEN

How Will We Know If It Worked? Evaluating Alt#4

The Alternative #4 business model is notionally new, and while based on reasonable economic principles (see Chapter Four), may require some adjustments to function optimally. The prototype of the Alt#4 business model is a demonstration/field experiment that, at the conclusion of the prototype period, will either be terminated or expanded. Effective evaluation can support both of these decisions.

Evaluation researchers (Clarke, 2005; Rossi, Lipsey, and Freeman, 2004) traditionally divide evaluation efforts into two groups: formative evaluations, whose objective is to support processes of program improvement, and summative evaluations, which aim to determine overall effectiveness of programs with an eye toward recommending whether they should continue. The Alt#4 prototype can benefit from both kinds of evaluation.

This chapter lays out an approach to evaluation of the Alt#4 prototype that will contribute to the (formative) improvement of the prototype as it progresses as well as make possible definitive (summative) judgments about the success or failure of the prototype at the conclusion of the three-year prototype period. The chapter proposes sets of metrics for three levels of evaluation: assessment of process and implementation, assessment of the outcome, and assessment of cost and efficiency. These three levels represent three of five levels from the hierarchy of evaluation, discussed in greater detail below.

The Hierarchy of Evaluation

Rossi, Lipsey, and Freeman (2004) specify five levels in the hierarchy of evaluation, each serving as a foundation for the levels above it (their hierarchy is duplicated in Figure 7.1). In the case of the Alt#4 prototype, the infamy of JSIMS preempted a formal needs assessment and predicated the TC AoA, pushing the process immediately to the second level of evaluation. The TC AoA did include efforts to define exactly *what* was needed, which is traditionally an important part of a needs assessment. The TC AoA began, and Chapter Four of this report completes, an assessment of design and theory for Alt#4. Alt#4 is based on sound economic principles, although experience and literature suggest that Alt#4 is not without risks. A prototype test of the business model appears to be a prudent way to proceed. Within the context of that prototype, three levels of assessment remain:

Figure 7.1
The Hierarchy of Evaluation

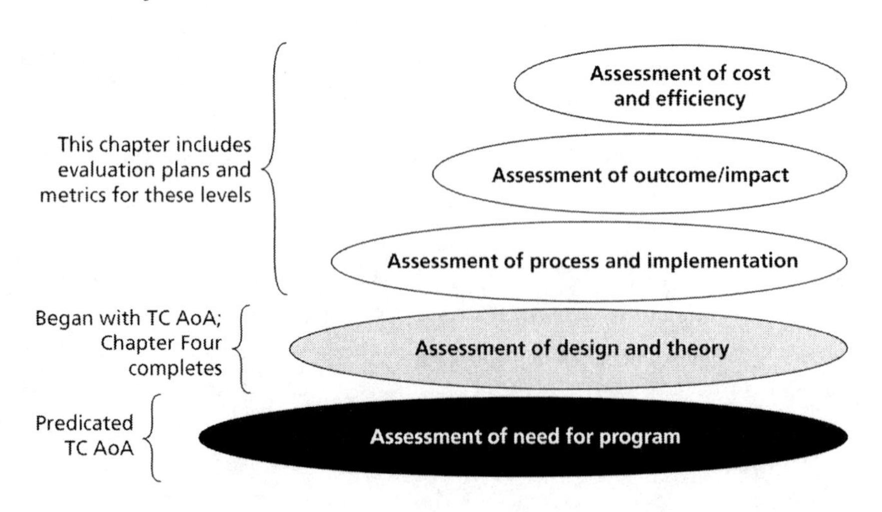

SOURCE: Based on Exhibit 3-C in Rossi, Lipsey, and Freeman (2004, p. 80).
RAND *MG442-7.1*

- Assessment of process and implementation: Is the prototype being done "right?" Are there things about the prototype which could be improved?
- Assessment of outcome: Is the prototype "working?"
- Assessment of cost and efficiency: If the prototype delivers what it is supposed to, is it doing so in a manner that realizes the efficiencies and cost savings it purports to?

The subsections below discuss these three levels of evaluation in greater detail and suggest criteria of merit (metrics) for each.

Implementation Evaluation

The assessment of process and implementation serves both formative and summative evaluation goals. From a formative standpoint, process evaluation can help identify areas for improvement in program processes or areas where aspects of the implementation are falling short of the business model's vision for them. From a summative standpoint, implementation assessment confirms that the program is actually an instance of what it is supposed to be, so as to deliver a fair verdict about the success of that type of program. In the Alt#4 prototype case, this is an assessment of the extent to which the Alt#4 business model is actually realized by the prototype, so at the end of the prototype its success or failure is appropriately attributable Alt#4. If the implementation assessment finds that the prototype is *not* a good instance of the Alt#4 business model, then the prototype is not a fair test of Alt#4. Both the summative and formative interest in the assessment of process and implementation hinges on whether the business model is being done "right" in the prototype.

Candidate assessment criteria at this level tend to be of one of two types: those that contribute to the assessment of the extent to which implemented processes *match* with the proposed business model and those that assess the extent to which those processes are actually *functioning*.

For the Alt#4 prototype, we propose several measures with which to assess the match between the prototype and the abstract business model:

- Were all model-specified components included in the prototype:
  - Is there a governance/oversight body specified?
  - Is there a catalog conductor?
  - Was an advisory board stood up?
  - Was a training user identified and engaged?
- Are all model-specified components active:
  - Has the advisory board met?
  - Have standards for catalog entry been established or adopted?
  - Have solicitations for training support been written?
  - Has seed money been spent?
- Do prototype component activities correspond with Alt#4 business model requirements:
  - Does training support solicited satisfy the learning objective selection criteria specified in Chapter Six?
  - Is the prototype executor directly making or buying simulation tools? If so, is this activity part of the prototype or affecting it in any way?
  - Is GFE being used for training tools under the prototype? Did it compete fairly to be used?
  - Do training solicitations and contracts comply with Alt#4 requirements (FFP, no TSP/tool vendor COI, tools only from catalog)?

If the prototype matches the business model, how can the DoD assess the extent to which implemented activities are functioning as intended? We propose the following metrics:

- number of training support solicitations written under the prototype;
- number of bids received for each training support solicitation;
- training support contracts written;

- number of training support contracts that comply with Alt#4 requirements (FFP, no TSP/tool vendor COI, tools only from catalog);
- number of solicitation written offering seed money;
- number of proposals received in response to solicitations offering seed money;
- number of seed money proposals funded;
- number of tools that pass compliance testing and enter toolbox/ catalog; and
- number of different vendors entering tools into catalog.

Clearly some of these measures also have implications at the outcome level of evaluation. Different levels in the evaluation hierarchy may rely on identical measures but consider them for different purposes. We now turn to the outcomes level.

Outcomes Evaluation

Evaluation at the outcome level is concerned with the actual results of the prototype: Did it "work?" Central in the assessment of outcomes is what Clarke (2005, p. 15) calls the program's "theory of effect"—how things are supposed to work. Chapter Two describes the logic of the Alt#4 model and includes its theory of effect. Alt#4 proposes to use a specific market structure to foster competition and innovation in the creation of tools and provision of training. If the process assessment for the prototype reveals that the implementation follows the standards for the Alt#4 business model, are the desired outcomes being realized? This section describes the incremental outcomes of the Alt#4 business model that are part of the model's theory of effect and suggest which data to collect to evaluate those outcomes.

Alt#4's theory of effect suggests that competition among tool vendors and between TSPs, if protected from perverse incentives resulting from certain DoD contract types and tool ownership structures, fosters innovation and allows for the provision of high-quality training support at good value. Although "at good value" is central at the top level of the evaluation hierarchy, assessment of cost and efficiency, competition, innovation, and provision of training are all targets of outcomes evaluations.

Provision of training is the central goal of any candidate business model in this area, so its consideration merits primacy. If training support is not being provided or is being provided inadequately, then something in the implementation or the business model needs to be changed (if considered formatively) or the prototype is a failure (if considered summatively). Assessment criteria for provision of training support can include:

- customer satisfaction with training support, both at training user management/command level and at the level of individual trainees; and
- if information about previous training of this task is available, is training reaching the same or a better standard than previous training in this area?

If the prototype is providing adequate training, then the evaluation can extend to cover supporting elements of the theory of effect, namely, competition in both markets and innovation in the tool marketplace.

Evidence of Competition

- number of different TSPs bidding on training support contracts;
- number of different TSPs winning and being awarded contracts;
- number of tools available in the tool catalog;
- number of different vendors placing tools in the catalog;
- number of different tools of a given type available in the catalog; and
- number of different tools in the catalog that are being used by TSPs.

Evidence of Innovation

- number of new tools entering the catalog;
- number of tools from vendors that received seed money support that enter the catalog;[1]
- number of new tools entering the catalog in areas specified for development attention by the catalog conductor, whether those tools received seed money stimulation or not; and
- surveys of training users (from satisfaction surveys) regarding innovation in the training support they received or in the tools used.

Evaluation of Efficiency

At the top of the evaluation hierarchy is the assessment of efficiency or cost. In addition to delivering quality training and innovative tools, the Alt#4 business model proposes to align incentives for cost efficiency and ensure cost efficiency through competition. If the outcomes assessment confirms that adequate training support is being delivered and that there is competition in both the training support market and the tool vendor market, how great are the cost savings realized by the DoD?

Cost benefit analyses are challenging in this case, especially if the training support purchased ends up being for a new training requirement (and no baseline against which to compare it). However, prototype executors should keep detailed cost data, and various cost metrics can be compared against external benchmarks as part of the assessment of efficiency. A good summary evaluand would be:

- aggregate dollar cost per training hour delivered.

---

[1] Mittal's (2005) congressional testimony regarding the ongoing success of the SBIR program notes that the GAO uses "successful commercialization" of products/firms supported with grants as a success metric.

There are other efficiency measures beyond training costs. There are also issues regarding the responsiveness of training to emergent needs and the efficiency of the catalog conductor's seed money investments. We suggest the following additional measures of efficiency:

- proportion of seed money awards that ultimately add a tool to the catalog; and
- time lag between specification and publication of a new training requirement and vendor/TSP ability to meet that requirement.

Doing Evaluation

When implementing a program prototype, there is a great deal to do, and collecting evaluation measurements may not always receive the highest priority. Because of the formative and summative benefits of evaluation in this context, it is critical that the prototype be evaluated. *We recommend that the DoD arrange for an impartial outsider to conduct the prototype evaluation.* Evaluation by a nonstakeholder decreases the likelihood that parochial interest will play a part in the evaluation and increases the legitimacy of the evaluation. Outside professionals with skill and experience in evaluation research will be more likely to produce an effective evaluation. Evaluation research is not free. Funding support for the evaluation of the prototype should be budgeted early on, either out of the prototype budget or through one of the primary stakeholders.

The best way to make sure that there are sufficient data for the evaluations is for the prototype executor to track data on critical transactions. Although evaluation is a core responsibility of the governance body, the actual action components (the catalog conductor, the contracting/grants support, and the training user) must carefully document their activities and expenditures. Even if critical data are just stored for future use, archived copies of all solicitations generated and all responses received (at all stages of the proposal and bid process) should be retained. When the catalog is established, catalog conductor personnel should make sure that the database in which it resides includes fields for date of certification and whether or not the vendor received seed money support from the catalog conductor or support from any DoD funding source (SBIR, etc.).

In addition to keeping careful records on the DoD component side, effective evaluation requires two active data collection efforts: first, satisfaction and performance surveys of the training users and, second, informational surveys of the training service providers. Information about the TSPs' experiences—which catalog tools they considered and which they ultimately used—and their understanding of prototype policies stand to make a significant contribution to the evaluation of the functioning of the prototype. Careful data tracking and plans for surveys of training users and TSPs should be added to the list of responsibilities presented in the implementation plan in Chapter Six.

CHAPTER EIGHT

Conclusions: The Challenges of Expanding Use of Alt#4 Beyond the Prototype

In this final chapter, we reiterate our key findings and conclusions and briefly consider the policy implications of a successful Alt#4 prototype.

Findings

The DoD undertook the TC AoA because of serious concerns about existing business models for the acquisition of simulations and simulation training support, typified by the failure of JSIMS. The TC AoA process produced an innovative acquisition alternative, Alt#4, which proposed a business model that would align the incentives of all participants toward the provision of responsive, high-quality training at best value prices. The design of the model requires that the DoD buy training support instead of tools *and* training support, buy that training support on FFP contracts, and ensure innovation and competition in the simulation tool market.

Chapter Three of this report presents case summaries for seven examples of simulation or simulation training acquisition. Four of the seven cases (VCCT, DMO, and the two examples from the British MOD NRTA) have features in common with the Alt#4 business model. These four cases clearly show that buying training or tools as a service on FFP contracts can lead to positive training outcomes and cost savings. The DMO case shows that models in which the DoD does not come to own the simulation tools can be effective even when simulations are heavyweight and highly military-specific in their use.

Chapter Four of this report considers the key features of the Alt#4 business model and some of the interesting differences in the other business models in light of theory and literature in the field of economics. The review of economic theory suggests that Alt#4 is based on reasonable and sound economic principles but that its implementation is not without risks or challenges.

Chapter Five identifies key challenges to a successful prototype and suggests mitigation strategies. Challenges include the difficulty of balancing compliance standards between exclusivity and permissiveness, finding a legal and effective way to invest seed money in the tool marketplace, writing solicitations for training that clearly express the training need while leaving providers enough flexibility to offer ways to meet that need, and identifying performance measures to ensure that training delivered under FFP contracts actually meets training needs. None of these risks is insurmountable but each will require care and attention on the part of the prototype executors.

Conclusion

Taken together, Chapters Three and Four suggest that the Alt#4 business model *could* work. This optimism is qualified by the presence of risks and challenges in theory and in practical experience and the observation that slightly different business models might be more appropriate for certain kinds of simulation or training acquisitions.

Recommendations

Our analysis leads us to recommend that the DoD proceed with the prototype of Alt#4. The observed balance between theoretical plausibility and empirical risks suggests that this activity is highly appropriate to a test, pilot, or prototype program implementation. We further recommend that the DoD strive to make the prototype implementation as close as possible to a test of the "pure" Alt#4 business model. Chapter Three shows that business models similar to Alt#4 can yield successes. The Alt#4 business model contains some innovative elements not seen in existing approaches, and the prototype is an important opportunity to see those innovations in action.

For the prototype to succeed and for it to be an accurate representation of the Alt#4 business model, certain elements must be in place and several risks avoided or mitigated. Chapter Six contains RAND's proposed implementation plan for the Alt#4 prototype. We recommend that the prototype executor establish four components for prototype execution: a governance body, a catalog conductor, an advisory board, and a prototype contracting/grants support. The prototype also needs to include a preexisting training user who is willing to buy training following the rules of the prototype.

Finally, to be useful, the prototype will need to be effectively evaluated. We recommend that the DoD arrange for an impartial outsider to conduct the prototype evaluation and that funding support for this effort be budgeted at the outset of the prototype. Evaluation should begin in the first year of the prototype so that constructive formative observations can lead to prototype adjustments as necessary.

Learning from the Prototype

Chapter Seven lists the elements of an evaluation plan for the prototype, including process evaluation, outcomes evaluation, and evaluation of efficiency. In this concluding section, we briefly consider the implications of the success or failure of the prototype.

If the Prototype Fails

Because there are several levels of evaluation under consideration (implementation, outcome, and efficiency—see Chapter Seven) the prototype could "fail" in several different places, each with different implications. If evaluation reveals that process and implementation fall short of what is planned, prototype executors can take immediate action to correct and improve their implementation. If, at the end of the prototype, the implementation is still found lacking, something more severe is implied: The prototype failed to implement Alt#4, and the outcomes and efficiency evaluations results pertain only to whatever actually was implemented. Failure to implement Alt#4 might suggest that either the business model is too challenging to actually implement with the resources committed to the prototype or the prototype executors failed to adhere to the model and instead implemented something else (a risk identified in Chapter Five).

If the process is successful, the prototype may still not realize desired outcomes and efficiencies. If that proves to be the case, the failure of the prototype would be less interesting than the reasons for that failure. Given that Alt#4 follows theoretically sound economic reasoning, how and in what ways the empirical situation had betrayed that reasoning could provide useful information for future simulation and training acquisitions.

If the Prototype Succeeds

Even if the Alt#4 prototype is an unqualified success, it may not be broadly adopted within the DoD. The biggest challenge facing expansion of the business model is policy inertia, and the possibility that relevant decisionmakers may not be aware of the alternative.

As the prototype proceeds, proponents must be strategic in announcing and communicating the success realized to raise awareness of the prototype and its potential. By the time the final prototype report announces the overall success or failure and lessons learned from the prototype, relevant high-level stakeholders should be eagerly awaiting the report and anticipating acting on its findings.

For this to occur, prototype proponents must engage in education and marketing activities relevant to the prototype, take steps to communicate the existence of the prototype, advertise progressive successes, and be prepared with an expansion plan at the (presumed) successful conclusion of the prototype period.

This education and communication plan should be aimed at a broadly defined array of stakeholders, not just high-level decisionmakers. If Alt#4 is to be expanded and be successful, it will require the support of the training user community. Training users play an absolutely critical role in each training provision transaction under the Alt#4 business model. If they are unwilling or unenthusiastic participants, an expansion may not realize the same successes as the prototype. However, if those in the training user community see and want Alt#4, they can move to adopt parts of the business model independently, hastening its expansion, and applying grass-roots pressure for expanded support and adoption at higher levels.

Finding the Best Way to Handle Venture-Capital-Like Investment Over a Broader Market

Chapter Six suggests that venture-capital-like activities for the prototype be conducted from within the DoD, in the catalog conductor component. If the prototype succeeds and application of the business model is expanded, it might be worth considering other options for organizing the disbursement of such investments, perhaps through a nonprofit venture-capital company external to the DoD, such as the CIA's In-Q-Tel or the U.S. Army's Venture Capital Initiative, OnPoint Technologies. Held et al. (2002) have a full discussion of an implementation strategy for a venture-capital corporation for the Army that would be a useful resource, should the DoD choose to head in that direction.

Domains in Which the Alt#4 Model Might Not Be Appropriate

Finally, a word of caution. Just because Alt#4 succeeds in a limited prototype context (assuming that it does), it might not be applicable to every simulation tool or training need. There may be certain simulation tool areas where the tool market is not sufficiently robust to support competition for tool provision, and the DoD would be better off with a slightly different business model. Also, in situations in which uniformed personnel provide all of the training and training users do not want to buy simulation training support as a service, perhaps the DMO model of buying simulator availability as a service would be most appropriate. Finally, Alt#4 is not appropriate where performance criteria cannot be specified and will struggle when performance criteria are poorly specified. Even if Alt#4 is wildly successful in the prototype, expanded application of the model should be thoughtful and careful and, when in doubt, should focus on Alt#4's core goals (aligning the incentives of all participants and buying training support and tools as a service) to get the right business model for the right individual circumstances.

Bibliography

Ashby, Jill, "Special Operation Forces Air Ground Interface Simulator (SAGIS)," briefing to the Milestone Decision Authority, January 22, 2004.

Bajari, Patrick, and Steve Tadelis, "Procurement Contracts: Fixed-Price vs. Cost-Plus," working paper, 1999.

———, "Incentives versus Transaction Costs: A Theory of Procurement Contracts," *RAND Journal of Economics*, Vol. 32, No. 3, 2001, pp. 287–307.

Baldwin, Laura H., Frank A. Camm, and Nancy Y. Moore, *Federal Contract Bundling: A Framework for Making and Justifying Decisions for Purchased Services*, Santa Monica, Calif.: RAND Corporation, MR-1224-AF, 2001.

Banerjee, Abhijit V., and Esther Duflo, "Reputation Effects and the Limits of Contracting: A Study of the Indian Software Industry," *The Quarterly Journal of Economics*, Vol. 115, No. 3, August 2000, pp. 989–1017.

Barco, "Barco System Solution Chosen for Navy MSAT Environment," press release, Norfolk, Va., 2005. Online at http://www.barco.com/corporate/en/pressreleases/show.asp?index=1465 (as of September 1, 2005).

BBN Technologies, "BBN Technologies and DARWARS Partners Redefine State-of-the-Art in Experience-Based Training," press release, Cambridge, Mass., December 6, 2004. Online at http://www.generalcatalyst.com/news/articles/bbn_041206.html (as of July 29, 2005).

———, "BBN Technologies Helps to Launch 'DARWARS Ambush!' New PC-Based Combat Team Trainer for U.S. Soldiers in Iraq," press release,

Cambridge, Mass., December 7, 2004. Online at http://www.bbn.com/News_and_Events/Press_Releases/04_12_07.html (as of July 29, 2005).

Berliner, Allison, "GSA Awards Contract Vehicle to 8(a) Companies," *Washington Technology*, June 7, 2004. Online at http://www.spangledesign.com/emcinc/GSAawardsContractVehicle.pdf (as of August 30, 2005).

Bilbruck, John, "Multi-Purpose Supporting Arms Trainer (MSAT) Program Overview," briefing, Orlando, Fla.: Surface and Expeditionary Warfare Programs, Naval Air Warfare Center Training Systems Division, 2004.

Birkler, J. L., Giles K. Smith, Glenn A. Kent, and Robert V. Johnson, *An Acquisition Strategy, Process, and Organization for Innovative Systems*, Santa Monica, Calif.: RAND Corporation, MR-1098-OSD, 2000.

Birkler, J. L., et al., *Assessing Competitive Strategies for the Joint Strike Fighter: Opportunities and Options*, Santa Monica, Calif.: RAND Corporation, MR-1362-OSD/JSF, 2001.

Birkler, J. L., et al., *Competition and Innovation in the U.S. Fixed-Wing Military Aircraft Industry*, Santa Monica, Calif.: RAND Corporation, MR-1656-OSD, 2003.

Bizub, Warren, "Training Capabilities Analysis of Alternatives Acquisition Prototype for Joint Close Air Support (JCAS)," U.S. JFCOM, n.d.

Blackmon, April, "War 'Games': Trailers House War Training," *Fort Riley Post*, Vol. 48, No. 43, October 28, 2005. Online at http://www.riley.army.mil/newspaper/index.htm (as of November 2, 2005).

Boeing Company, "Boeing Opens F-15C Distributed Mission Training Facility in Japan," press release, May 3, 2005. Online at http://www.boeing.com/news/releases/2005/q2/nr_050503m.html (as of August 1, 2005).

Brower, J. Michael, "Distributed Mission Training," *Military Training Technology Online*, Vol. 8, No. 4, November 19, 2003. Online at http://www.military-training-technology.com/article.cfm?DocID=272 (as of August 1, 2005).

Business Executives for National Security, *Accelerating the Acquisition and Implementation of New Technologies for Intelligence: The Report of the In-*

dependent Panel on the Central Intelligence Agency In-Q-Tel Venture, Washington, D.C., June 2001.

Carney, David J., and Patricia A. Oberndorf, "The Commandments of COTS: Still in Search of the Promised Land," *The Journal of Defense Software Engineering,* May 1997.

Clarke, Alan, *Evaluation Research: An Introduction to Principles, Methods and Practice,* London: SAGE Publications Ltd., 2005.

Clark, COL Julius E., "Army Joint Support Team," briefing, April 2005.

Conduct Close Air Support, "Joint Tactical Task," JC meeting, JTRAT Approved, Data Base References (Tactical 3.2.2), May 21, 2005.

"Consortia Invited to Begin Negotiations," *Preview,* April 2005.

Crocker, K. J., and K. J. Reynolds, "The Efficiency of Incomplete Contracts: An Empirical Analysis of Air Force Engine Procurement," *RAND Journal of Economics,* Vol. 24, 1993, pp. 126–146.

Davis, Paul K., and Robert H. Anderson, *Improving the Composability of Department of Defense Models and Simulations,* Santa Monica, Calif.: RAND Corporation, MG-101-OSD, 2003.

Defense Advanced Research Projects Agency, "About DARWARS," 2005, Online at http://www.darwars.com/about/index.html (as of July 29, 2005).

———, "DARPA'S DARWARS Program Debuts at I/ITSEC Show with Ten Major Training Technology Participants," Orlando, Fla., December 1, 2004. Online at http://www.darwars.com/downloads/Press%20Release.doc (as of July 29, 2005).

———, "DARWARS at I/ITSEC 2004," fact sheet. Online at http://www.darwars.com/downloads/Ten_Vendors.pdf (as of July 29, 2005).

Defense Grant and Agreement Regulatory System, "Other Transactions." Online at http://alpha.lmi.org/dodgars/other_transactions/other_transactions.htm (as of August 23, 2005).

———,"Technology Investment Agreement (TIA)." Online at http://alpha.lmi.org/dodgars/tias/tias.htm (as of August 23, 2005).

Defense Modeling and Simulation Office, *High Level Architecture for Modeling and Simulation Management Plan,* Version 1.6, July 1995.

Department of National Defence, *The Joint Simulation and Modeling for Analysis, Requirements, Training, and Support (SMARTS) Initiative: A Vision for Enabling Strategy 2020 Through the Application of Modeling & Simulation in DND*, Canadian Forces report, Canada, March 31, 2004.

Department of the Navy, Acquisition One Source, "Other Transactions." Online at http://navyaos.ati4it.com/navyaos/content/view/full/136 (as of August 22, 2005).

Director of Defense Research and Engineering, "Revision 1 to Guidance on Instruments for Stimulation or Support of Research," Memorandum for Secretaries of the Military Departments, Washington, D.C., March 24, 1998.

Drezner, Jeffrey A., and Robert S. Leonard, *Global Hawk and DarkStar: Their Advanced Concept Technology Demonstration Program Experience*, Santa Monica, Calif.: RAND Corporation, MR-1473, 2002.

Erwin, Sandra I., "F/A-22 Pilots Begin Training at Tyndall AFB," *National Defense Magazine*, November 2003. Online at http://www.nationaldefensemagazine.org/issues/2003/Nov/Planned_upgrades.htm (as of August 30, 2005).

Farrell, Joseph, and Paul Klemperer, "Coordination and Lock-In: Competition with Switching Costs and Network Effects," working paper, December 2004.

FATS, Inc., "FATS, Inc. Teams with Lockheed Martin to Build Reconfigurable Vehicle Simulator for U.S. Army," press release, June 21, 2005. Online at http://www.fatsinc.com/about/news/pr80.cfm (as of July 29, 2005).

Foliente, Rodney, "4ID Virtually Trained for Iraq," *4th Infantry Division News*. Online at http://www.hood.army.mil/4id/News/Archive/2005/VIRTUALCONVOY.html (as of November 2, 2005).

Gansler, Jacques S., "Moving Toward Market-Based Government: The Changing Role of Government as the Provider," IBM Endowment for the Business of Government, June 2003. Online at http://www.businessofgovernment.org/pdfs/Gansler_Report.pdf (as of June 2003).

Garrabrants, William, et al., "Novel Business Model Approach for Future JSIMS Acquisition," Paper No. 1876 presented at the Interservice/Industry Training, Simulation, and Education Conference, Orlando, Florida, 2004.

Gholz, Eugene, "MOSA II Business Model," paper presented at the Aging Aircraft 2005 convention, Palm Springs, Calif. Online at http://www.jcaa.us/AA_Conference2005/Avionics/Ses10/10_06_Gholz.pdf (as of August 18, 2005).

Gibbons, Robert, "Incentives in Organizations," *The Journal of Economic Perspectives*, Vol. 12, No. 4, Autumn 1998, pp. 115–132.

Gourley, Scott R., "Training for the Ambush," *Military Training Technology*, Vol. 9, No. 5, October 27, 2004. Online at http://www.military-training-technology.com/article.cfm?DocID=663 (as of July 29, 2005).

"Government Drops Army Tank Training Deal," *YAHOO!News*. Online at http://uk.news.yahoo.com/050615/325/fl9e8.html (as of June 15, 2005).

Griffin, Sean P., Ernest H. Page, Zachary Furness, and Mary C. Fischer, "Providing Uninterrupted Training to the Joint Training Confederation (ITC) Audience During the Transition to the High Level Architecture (HLA)," *Proceedings of the 1997 Simulation Technology and Training Conference*, Canberra, Australia, March 17–20, 1997.

Hamilton, Laura, Brian M. Stecher, and Stephen P. Klein, *Making Sense of Test-Based Accountability in Education*, Santa Monica, Calif.: RAND Corporation, MR-1554-EDU, 2002.

Held, Bruce J., et al., *Seeking Nontraditional Approaches to Collaborating and Partnering with Industry*, Santa Monica, Calif.: RAND Corporation, MR-1401-A, 2002.

Held, Bruce J., and Ike Yi Chang, *Using Venture Capital to Improve Army Research and Development*, Santa Monica, Calif.: RAND Corporation, IP-199, 2000.

Her Majesty's Treasury, "Public Private Partnerships," n.d. Online at http://www.hm-treasury.gov.uk/documents/public_private_partnerships/ppp_index.cfm (as of September 13, 2005).

Her Majesty's Treasury and Controller of Her Majesty's Stationery Office, *PFI: Meeting the Investment Challenge*, London: July 2003.

Holmstrom, Bengt, and John Roberts, "The Boundaries of the Firm Revisited," *Journal of Economic Perspectives*, Vol. 12, No. 4, Fall 1998, pp. 73–94.

In-Q-Tel, "BENS Panel Says In-Q-Tel Model Makes Good Business Sense," press release, Washington, D.C., August 7, 2001. Online at http://www.in-q-tel.com/news/releases/08_07_01b.html (as of August 30, 2005).

———, "About Us: Model," 2005a. Online at http://www.in-q-tel.org/about/model.html (as of August 30, 2005).

———, "Investing in Our Nation's Security," 2005b. Online at http://www.in-q-tel.org/about/index.htm (as of August 30, 2005).

———, "Strategic Investments, Targeted Returns," 2005c. Online at http://www.in-q-tel.org/invest/index.htm (as of August 30, 2005).

JFCOM JWFC/JNTC Innovative Acquisition Strategy Offsite, August 3, 2005, BMH conference room, Suffolk, Va.

Joint Tactics, Techniques, and Procedures for Close Air Support (CAS), Joint Publication 3-09.3, Washington, D.C., September 3, 2003.

Joskow, Paul, "Contract Duration and Relationship-Specific Investments: Empirical Evidence from Coal Markets," *The American Economic Review*, Vol. 77, No. 1, March 1987, pp. 168–185.

Katz, Warren, *Psychological Dynamics of the CPFF Business Model*, MAK Technologies, April 17, 2002.

Kettl, Donald F., *Sharing Power: Public Governance and Private Markets*, Washington, D.C.: The Brookings Institution, 1993.

Klerman, Jacob Alex, "Measuring Performance," in Robert E. Klitgaard et al., eds., *High-Performance Government: Structure, Leadership, Incentives*, Santa Monica, Calif.: RAND Corporation, MG-256-PRGS, 2005, pp. 343–379.

Kortum, Samuel, and Josh Lerner, "Does Venture Capital Spur Innovation," Cambridge, Mass.: National Bureau of Economic Research, working paper 6846, December 1998. Online at http://www.nber.org/papers/w6846 (as of August 30, 2005).

Levin, R. E., Director, Acquisition and Sourcing Management, "Defense Acquisitions: Incentives and Pressures That Drive Problems Affecting Satellite and Related Acquisitions," letter to The Honorable C. W. Bill Young, Chairman, Subcommittee on Defense, Committee on Appropriations, House of Representatives, Washington, D.C., GAO-05-570R Space Systems Acquisitions, June 23, 2005.

Lockheed Martin, "Virtual Combat Convoy Trainer: Description," Online at http://www.lockheedmartin.com/wms/findPage.do?dsp=fec&ci=15346&rsbci=0&fti=126&ti=0&sc=400 (as of July 29, 2005).

"Lockheed Wins $4.2M Trainer Contract," *Orlando Business Journal*, June 27, 2005. Online at http://www.bizjournals.com/orlando/stories/2005/06/27/daily2.html?f=et70 (as of August 30, 2005).

Lorell, Mark A., and John C. Graser, *An Overview of Acquisition Reform Cost Savings Estimates*, Santa Monica, Calif.: RAND Corporation, MR-1329-AF, 2001.

MacDonald, Elizabeth, and Robert Langreth, "Spore Wars," *Forbes Magazine*, June 6, 2005.

McKaughan, Jeff, "Special Operators, Special Tactics," *Special Operations Technology*, July 23, 2003. Online at http://www.special-operations-technology.com/article.cfm?DocID=158 (as of August 30, 2005).

MetaVR, "Case Studies: MetaVR Visuals Used in III Corps Convoy Simulation Training." Online at http://www.metavr.com/casestudies/convoytrainer.html (as of July 29, 2005).

Mittal, Anu K., "Testimony Before the Subcommittee on Environment, Technology, and Standards, Committee on Science, House of Representatives: Observations on the Small Business Innovation Research Program," Washington, D.C., GAO-05-861T, June 28, 2005. Online at http://www.gao.gov/new.items/d05861t.pdf (as of August 24, 2005).

Murray, Bill, "GSA Seeks FAST Streamlining, Savings," *Government Computer News*, Vol. 17, No. 10, April 27, 1998. Online at http://www.gcn.com/17_10/news/33363-1.html (as of August 30, 2005).

Nash, Major General Gordon C., USMC Commander, Joint Warfighting Center and Director for Joint Training, U.S. Joint Forces Command, "Statement Before the House Armed Services Subcommittees on Readiness Terrorism, Unconventional Threats and Capabilities on the Joint National Training Capability," March 18, 2004. Online at http://www.house.gov/hasc/openingstatementsandpressreleases/108thcongress/04-03-18nash.pdf (as of July 29, 2005).

Naval Air Systems Command Training Systems Division, *Statement of Work for Multi-Purpose Supporting Arms Trainer: Phase II Development*, Orlando, Fla., July 14, 2005.

Network Centric Operations Industry Consortium, "An Introduction to the Network Centric Operations Industry Consortium (NCOIC)," position paper V2.0, March 2005.

Nurse, Charles, "Innovative Use of Other Transactions (OT)," Space and Naval Warfare Systems Command, Navy Acquisition Reform Senior Oversight Council (NARSOC) briefing, January 29, 1998. Online at http://www.abm.rda.hq.navy.mil/navyaos/content/download/1520/7535/file/midsnurs.pdf (as of August 22, 2005).

O'Hara, Terence, "In-Q-Tel, CIA's Venture Arm, Invests in Secrets," *Washington Post*, August, 15, 2005. Online at http://www.washingtonpost.com/wp-dyn/content/article/2005/08/14/AR2005081401108_pf.html (as of August 19, 2005).

Office of Secretary of Defense (OSD) and United States Joint Forces Command (USJFCOM), *Training Capabilities Analysis of Alternatives (TC AoA) Final Report*, Vol. I, Washington, D.C., July 30, 2004.

"Prepared to Play," *Training and Simulation Journal*, October 1, 2001. Online at http://www.tsjonline.com/story.php?F=340385 (as of August 30, 2005).

"Pricing Models," Infotechnet.org. Online at http://www.infotechnet.org/ntca/BusinessModels.htm (as of August 22, 2005).

Qualters, Sheri, "Small Defense Contractors Unite in Bid for More Work," *Boston Business Journal*, January 14, 2005. Online at http://www.bizjournals.com/boston/stories/2005/01/17/story8.html?t=printable (as of August 18, 2005).

Rietze, Susan, "Distributed Mission Ops Shape USAF Training Projects," *National Defense Magazine*, November 2003. Online at http://www.nationaldefensemagazine.org/issues/2003/Nov/Distributed_Mission.htm (as of August 1, 2005).

Rodrigues, Louis J., Testimony Before the Subcommittee on Readiness and Management Support, Committee on Armed Services, U.S. Senate, "Defense Acquisition, Best Commercial Practices Can Improve Program Outcomes," Washington, D.C., GAO/T/NSZID-99-116, March 17, 1999.

Rogerson, William, "Profit Regulation of Defense Contractors and Prizes for Innovation," *Journal of Political Economy*, Vol. 97, No. 6, 1989, pp. 1284–1305.

Root, Lawrence M., Jerry Osterheld, and Mark McAuliffe, "Development Baton Handoffs," SimVentions White Paper, n.d. Online at http://www.simventions.com/whitepapers/03F-SIW-013.pdf (as of July 29, 2005).

Rossi, Peter H., Mark W. Lipsey, and Howard E. Freeman, *Evaluation: A Systematic Approach*, Thousand Oaks, Calif.: SAGE Publications, 2004.

Schank, John F., et al., *Options for Reducing Costs in the United Kingdom's Future Aircraft Carrier (CVF) Programme*, Santa Monica, Calif.: RAND Corporation, MG-240-MOD, 2005.

Schmidt, Conrad Peter, *Changing Bureaucratic Behavior: Acquisition Reform in the United States Army*, Santa Monica, Calif.: RAND Corporation, MR-1094-A, 2000.

Scott, Denise C., "Other Transactions," PowerPoint presentation, Research Development and Engineering Command–Armament Research Development and Engineering Center Legal (RDECOM-ARDEC LEGAL), May 25, 2005. Online at http://www.jhuapl.edu/aboutapl/events/industry2005/pdf/Scott%20-%20RDECOMota.pdf (as of August 23, 2005).

Slabodkin, Gregory, "DoD Integrates Best Sim Tools," *Government Computing News*, Vol. 16, No. 27, September 15, 1997. Online at http://appserv.gcn.com/16_27/news/32192-1.html (as of July 29, 2005).

"Special Operations Air-Ground Interface Simulator (SAGIS)—AFSOC," draft, October 17, 2003.

Stecher, Brian M., and Sheila I. Barron, *Quadrennial Milepost Accountability Testing in Kentucky*, Los Angeles: National Center for Research on Evaluation, Standards, and Student Testing, CSE Technical Report 505, 1999.

Stephenson, Daryl, "Training Centers Provide Battlefield Realism," *All Systems Go: Journal of Boeing Integrated Defense Systems*, Vol. 1, No. 7, pp. 8–9. Online at http://www.mdc.com/ids/allsystemsgo/issues/vol1/num7/issue7spreads.pdf (as of August 30, 2005).

Strategypage.com, "WARSIM Wobbles into Action," *Wargame News*, Vol. 3, February 2005. Online at http://www.strategypage.com/messageboards/messages/564-12.asp (as of July 29, 2005).

"Supplement to 1994 Interim Guidance for 10 U.S.C. §2371 Revision 1/ March 3, 1998," Guidance on 'Technology Investment Agreements' for Military Departments and the Defense Advanced Research Projects Agency, Washington, D.C.

Taylor, Curtis, and Steven Wiggins, "Competition or Compensation: Supplier Incentives Under the American and Japanese Subcontracting Systems," *The American Economic Review*, Vol. 87, No. 4, September 1997, pp. 598–618.

"The I-Fact of the Matter," *Defence Today*, n.d.

"The Joint Training System: A Primer for Senior Leaders," Chairman Joint Chiefs of Staff (CJCS), Guide 3501, Washington, D.C., October 10, 2003.

Tirole, Jean, *The Theory of Industrial Organization*, Cambridge, Mass.: The MIT Press, 1997.

_______, "The Analysis of Tying Cases: A Primer," *Competition Policy International*, Vol. 1, No. 1, Spring 2005.

Tiron, Roxana, "Pentagon Cancels Program with 'Checkered' Past," *National Defense Magazine*, April 2003. Online at http://www.nationaldefensemagazine.org/issues/2003/apr/Pentagon_Cancels.htm (as of July 29, 2005).

"Training and Mission Rehearsal Capabilities Solutions Proposal," Macrosystems, PowerPoint presentation, n.d.

U.S. Air Force, "Distributed Mission Operations Center (DMOC)," 2005. Online at http://www.dmoc.kirtland.af.mil/history/history_2.htm (as of August 30, 2005).

U.S. Army Communications-Electronics Command, "United States Army Soldier and Biological Chemical Command Broad Agency Announcement," Solicitation Number DAAB07-02-R-B-223, August 29, 2002.

U.S. Code, Title 5, "Government Organization and Employees," Appendix, "Federal Advisory Committee Act Amendments," Washington, D.C.

U.S. Department of Defense, "Contracts," news release, No. 066-99, February 18, 1999. Online at http://www.defenselink.mil/contracts/1999/c02181999_ct066-99.html (as of August 30, 2005).

———, "Contracts," news release, No.572-00 September 18, 2000. Online at http://www.defenselink.mil/contracts/2000/c09182000_ct572-00.html (as of August 30, 2005).

———, "'Other Transactions' (OT) Guide for Prototype Projects," January 2001. Online at http://www.afmc-pub.wpafb.af.mil/HQ-AFMC/PK/pkt/OTGuideAug2002.doc (as of August 24, 2005).

———, "Contracts," news release, No. 394-01, August 24, 2001. Online at http://www.defenselink.mil/contracts/2001/c08242001_ct394-01.html (as of August 30, 2005).

———, "Contracts," news release, No. 036-03, January 24, 2003. Online at http://www.defenselink.mil/contracts/2003/c01242003_ct036-03.html (as of August 30, 2005).

———, "Contracts," No. 917-03, December 5, 2003. Online at http://www.defenselink.mil/contracts/2003/ct20031205.html (as of August 30, 2005).

———, *Department of Defense Federal Advisory Committee Management Program*, Department of Defense Directive 5105.4, Washington, D.C., 2003.

———, "Contracts," news release, No. 314-05, April 4, 2005. Online at http://www.defenselink.mil/contracts/2005/ct20050404.html (as of August 30, 2005).

———, "Contracts," No. 535-05, May 31, 2005. Online at http://www.defenselink.mil/contracts/2005/ct20050531.html (as of August 30, 2005).

———, "Public, M&S Resources: Online M&S Glossary" (DoD 5000.59-M). Online at https://www.dmso.mil/public/resources/glossary/results?do=get&def=297, (as of September 27, 2005).

U.S. General Accounting Office, *Military Readiness: Lingering Training and Equipment Issues Hamper Air Support of Ground Forces*, Report to the Ranking Minority Members, Subcommittees on Total Force and Readiness, Committee on Armed Services, House of Representatives, GAO-03-505, Washington, D.C., May 2003.

———, "Report to the Chairman and Ranking Minority Member, Committee on Armed Services, U.S. Senate, Defense Acquisitions: DOD Has Implemented Section 845 Recommendations but Reporting Can Be Enhanced," GAO-03-150, Washington, D.C., October 2002. Online at http://www.gao.gov/new.items/d03150.pdf (as of August 23, 2005).

U.S. General Services Administration, "8(a) Federal Acquisition Services for Technology." Online at http://www.gsa.gov/Portal/gsa/ep/channelView.do?pageTypeId=8199&channelPage=%252Fep%252Fchannel%252FgsaOverview.jsp&channelId=-13469 (as of August 30, 2005).

———, "Small Business Governmentwide Acquisition Contracts Center," n.d. Online at http://www.gsa.gov/Portal/gsa/ep/channelView.do?pageTypeId=8199&channelPage=%252Fep%252Fchannel%252FgsaOverview.jsp&channelId=-13266 (as of August 30, 2005).

U.S. Government Accountability Office, *Report to Congressional Committees, Homeland Security: Further Action Needed to Promote Successful Use of Special DHS Acquisition Authority*, GAO-05-136, Washington, D.C., December 2004.

———, *Military Training: Actions Needed to Enhance DOD's Program to Transform Joint Training*, GAO-05-548, Washington, D.C., June 2005a.

———, *Report to the Subcommittee on Readiness and Management Support, Committee on Armed Services U.S. Senate: Defense Management, DoD Needs to Demonstrate That Performance-Based Logistics Contracts Are Achieving Expected Benefits*, GAO-05-966, Washington, D.C., September 2005b.

U.S. Securities and Exchange Commission, "Market Maker," March 17, 2000. Online at http://www.sec.gov/answers/mktmaker.htm (as of August 24, 2005).

U.S. Special Operations Command, Office of the Deputy Commander, *Joint Operational Requirements Document for the SOF Air Ground Interface Simulator (SAGIS): ACAT Level III*, MacDill Air Force Base, Fla., n.d.

Walker, Karen, "Army Eyes Convoy Simulator Expansion," *Training and Simulation Journal*, December 1, 2004a. Online at http://tsj.dnmediagroup.com/story.php?F=563287 (as of July 29, 2005).

———, "Pack Mentality: Plug-and-Play Simulators That Work Together Allow Team Training," *Training and Simulation Journal*, December 1, 2004b. Online at http://tsj.dnmediagroup.com/story.php?F=563303 (as of August 30, 2005).

Williamson, Oliver E., "Assessing Vertical Market Restrictions: Antitrust Ramifications of the Transaction Cost Approach," *University of Pennsylvania Law Review*, Vol. 127, 1979, pp. 953–993.

———, "The Limits of the Firm: Incentive and Bureaucratic Features," *Transaction Cost Economics*, Vol. I, Brookeld, Vt.: Edward Elgar Publishing Limited, 1985.

Wyld, David C., *The Auction Model: How the Public Sector Can Leverage the Power of E-Commerce Through Dynamic Pricing*, Grant Report for The PricewaterhouseCoopers Endowment for The Business of Government, October 2000.

Wysocki, Bernard, Jr., "U.S. Struggles for Drugs to Counter Biological Threats; As Bigger Firms Shun Effort, Small Ones Are Challenged; 'This Is Really Hard Stuff,'" *Wall Street Journal (Eastern Edition)*, July 11, 2005, p. A1.

Yannuzzi, Rick E., "In-Q-Tel: A New Partnership Between the CIA and the Private Sector," *Defense Intelligence Journal*, Vol. 9, No. 1, Winter 2000. Online at http://www.cia.gov/cia/publications/inqtel/ (as of August 17, 2005).